Edizioni PensareDiverso
Collana Cenacolo Jung Pauli.

A whisper. A heartbeat. A flash of awareness.
It is everywhere. It is everything.
To enter this book is to knock on the door of an ancient concept,
and allowing oneself to be led beyond the threshold of reason.

Bruno Del Medico

From Spinoza to Quantum Consciousness.

"Deus sive Natura." Spinoza's thought as a precursor to the current unified view of the cosmos. Modern panpsychism.

Bruno Del Medico Publisher.
PensareDiverso Editions
edizioni@delmedico.it
www.giardinoquantico.it

Summary

Introduction

For centuries, humanity has pondered the mystery of consciousness and the relationship between mind and matter. Some have sought answers in the depths of religion, others in the abstraction of philosophy, and still others in the increasingly sophisticated tools of science. But there is one idea that spans the ages and seems to resurface periodically with renewed vigor: the possibility that consciousness, or at least a primitive form of experience, is a fundamental and omnipresent property of the universe. This primordial idea resurfaces powerfully in the debate currently involving neuroscientists and philosophers of consciousness. This debate, born in the wake of new quantum physics, has developed with such impetus that it has also attracted the interest of the general public.

Spinoza's concept of "Deus sive Natura" represents one of the most powerful and radical formulations of this vision: everything that exists is an expression of a single divine-natural substance, and what we call "spirit" or "consciousness" is not a rare accident, but the very fabric of being. From this perspective, the starry sky, the beating of a cell, the motion of subatomic particles, and the intuition of a poet are not separate phenomena, but manifestations of a vital and conscious continuum.

This book embarks on a journey from archaic thought to contemporary scientific debate, traversing the thinking of ancient philosophers, medieval mystics, modern idealists, and quantum mind theorists. The aim is to reexamine Spinozian thought beyond its historical dimension, initiating a reflection on its implications for the future of knowledge. This exploration will not be limited to description, but will seek to bring different traditions and disciplines into dialogue: philosophy, physics, neuroscience, and ethics.

The question that accompanies the narrative is simple and at the same time dizzying: what if consciousness were not a late product of evolution, but an original property of reality itself? If this were true, every fragment of the universe—from stone to thought—would participate in the same mysterious awareness. An awareness that unites us not only with other human beings, but with the entire cosmos.

Throughout its chapters, the book illustrates how this idea was born, transformed, criticized, and is still being revived today by scientists and philosophers. It is not a question of analyzing a dogma, but of opening up a space for reflection capable of overcoming the barriers between disciplines and worldviews. Because, perhaps, understanding consciousness means rethinking the entire fabric of reality—and perhaps, ultimately, it means rethinking ourselves.

I. r God as Nature.

There are ideas that, although born centuries ago, continue to dialogue with our present. Baruch Spinoza's "Deus sive Natura" is one of them. The expression is as simple as it is radical: God is Nature. Not an anthropomorphic God, not an external will that moves the world from above, but the universe itself, with its laws, its structure, its intrinsic intelligence. When Spinoza published "*Ethica, ordine geometrico* demonstrata" in 1677, his equation of God with Nature caused scandal. For the theology of the time, it was almost sacrilegious. Yet this insight—indivisible and monistic—would take deep root in modern philosophy, even entering into dialogue with neuroscience and contemporary physics.

Today, talking about panpsychism often means returning to that starting point. The idea that every fragment of reality possesses a mental aspect, a form of consciousness or pre-consciousness, is not merely metaphysical: it is a bridge between physics and philosophy, between science and poetry. It is not surprising that Einstein, an attentive reader of Spinoza, declared his belief in "*a God who reveals himself in the harmony of all that exists, not in a God who concerns himself with the fate and actions of men.*"

The first part of this chapter will explain why an essay on panpsychism is necessary today. The reasons are philosophical, but also scientific and cultural. We live in an age in which artificial intelligence is forcing us to redefine the concept of consciousness. At the same time, advances in quantum physics seem to open up avenues towards an idea of the universe as a living organism. Understanding *Deus sive Natura* means rethinking the way we conceive of human beings within the cosmos.

Defining this Spinozian concept will be the second step. We will recount Spinoza's vision without reducing it to abstract formulas, remembering the context of his life: his exile from the Jewish community of Amsterdam, his scholarly solitude in small Dutch villages, his friendship with thinkers such as Henry

Oldenburg. Philosophy is not born in a vacuum, but in the tension between thought and the wounds of living.

We will then talk about what panpsychism is and how it differs from other conceptions of the mind. It is not naive animism, nor pure poetic metaphor; it is a theory that takes seriously the hypothesis that consciousness—or at least a proto-conscious core—is a universal attribute of matter. This perspective, already present in the pre-Socratics and Plotinus, passes through Giordano Bruno's Renaissance and reaches the contemporary scientist who wonders whether an electron can somehow 'feel' its own state.

The section dedicated to the attraction to the living cosmos will show the existential side of this idea. This is not just speculation: it is an invitation to perceive reality as part of a living fabric, of which we are temporary nodes. From Lucretius to Whitman, poets and thinkers have celebrated this vision, transforming it into an aesthetic and spiritual experience.

Finally, the common thread of the work—from the universal spirit to neuroscience—will lead the reader into the territory where ancient thought meets the most advanced science. Quantum consciousness, integrated information theories, the hypothesis that the brain and matter share fundamental structures: these are all elements that, like it or not, bring us back to that luminous formula conceived by a Dutch philosopher four centuries ago.

If the universe is alive and conscious in every part, then *Deus sive Natura* is not just a philosophical concept: it is the most sincere and dizzying description of our place in the whole.

Why an essay on panpsychism today.

Panpsychism is a philosophical position according to which consciousness, or at least an elementary form of subjective experience, is a fundamental and omnipresent property of reality. Consciousness would be present in some way in all entities, even the simplest ones such as particles or physical fields.

There are ideas that recur cyclically, like underground currents that resurface in times of cultural crisis. Panpsychism is one of them. It is not just an ancient philosophical concept: it is a key to understanding our time, suspended between the hypothesis of a mechanical universe and the growing intuition that matter hides a living dimension.

There are many reasons for writing an essay on this topic today. The philosophy of panpsychism, which affirms the presence of a mental aspect in everything, comes at a time when consciousness has returned to the center of the debate. Neuroscience seeks to define the boundaries of the mind. Artificial intelligence forces us to ask ourselves whether consciousness is a biological privilege or an emergent property. Theoretical physics considers, at least in some of its interpretations, that information and consciousness are not accidental by-products, but constituent elements of the cosmos.

For today's reader, panpsychism is not outdated mysticism. It is a way of thinking that dialogues with science. In 1929, British physicist Sir Arthur Eddington wrote that "*the matter of the physical world is only the shadow of something more real.*" This statement does not belong to a mystic, but to a scientist in the midst of the positivist era. Today, scholars such as Philip Goff take up that insight with rigorous arguments, analyzing whether consciousness can be an intrinsic property of matter, like mass or electric charge.

Contemporary culture, fragmented between hypertechnology and the search for meaning, finds in panpsychism a bridge between two worlds. In a NASA laboratory, the limits of artificial intelligence applied to Martian Rover technology are being discussed, while in the classrooms of Oxford, the concept of subjective experience in the simplest organisms is being re-examined. This parallelism shows that the issue is not marginal: it concerns both our ability to recognize life and the way we relate to what we do not immediately perceive as 'conscious'.

From a cultural point of view, panpsychism responds to the weariness of the concept of reductionism. After centuries in which science has broken down reality into ever smaller elements, many scientists and philosophers feel the need to recompose a unified picture. This is where the power of Deus sive Natura comes in. Spinoza's vision, which unites God and the universe, can now be read in the light of information theory and quantum physics. The idea that every phenomenon has an inner aspect resonates with hypotheses that consciousness may be diffused throughout the cosmos but manifests itself in different forms.

Popular culture also feels this tension. In the science fiction novels of Ursula K. Le Guin, in the films of Terrence Malick, in the essays of Carlo Rovelli, the need to perceive the cosmos as a living reality emerges. This book was written to provide a historical and scientific framework for that perception. Not to transform it into a new religion, but to make it a field of investigation.

Talking about panpsychism today therefore means talking about the nature of reality, the place of consciousness, the meaning of our existence. It means resuming a conversation that began with Anaximander, passed through Spinoza, was enriched by William James and Alfred North Whitehead, and is now expressed in the conferences of neuroscientists and physicists. It is a theme that concerns thought, science, and culture.

Defining 'Deus sive Natura'.

When Baruch Spinoza wrote Deus sive Natura (God or Nature), he made a bold move. He did not invent a simple aphorism, but formulated an entire worldview. For him, God was not a distant ruler, nor a creator who regulated the universe from outside. God was the universe itself with all its processes, immutable in its laws but changeable in its forms.

To understand the significance of this phrase, one must imagine 17th-century Holland. Amsterdam is a lively and cosmopolitan city, a crossroads of commerce and ideas. But it was also a place where theology maintained tight control. Spinoza, born in Amsterdam in 1632, was the son of a Sephardic Jewish family originally from the Iberian Peninsula (Portugal), who had taken refuge in the Netherlands to escape religious persecution by the Inquisition. Spinoza grew up in the Sephardic Jewish community, surrounded by sacred books and philosophical discussions. In 1656, he was expelled with a cherem, one of the most severe forms of excommunication, because his thinking was considered heretical. He abandoned the name Bento and retired to the villages of Rijnsburg and Voorburg, where he devoted himself to writing. It is necessary to clarify his names.

Bento de Spinoza was his birth name in Portuguese, used in the Sephardic community of Amsterdam, which excommunicated him. Baruch Spinoza was the Hebrew version of his name, literally 'blessed', which he adopted in religious and intellectual contexts. In Latin, in his philosophical works, he used the form Benedictus de Spinoza.

His *Ethics* (abbreviation of the original Latin name of the work: "*Ethica, ordine geometrico* demonstrata" published posthumously in 1677) is a geometric work: each idea is demonstrated as a theorem. In Part I, Spinoza declares that r God is the infinite, eternal substance of which everything is a manifestation. Nature is not a creature of God, but is God himself. (In Latin, *Deus sive Natura)*. It is a radical monism that rejects the separation between spirit and matter.

This concept has profound implications. If all that exists is God, then there is no place for miracles understood as a suspension of natural laws: the laws are God's own actions. There is no hierarchy between man, animal, stone, or star: all are modes of the same substance. It is a vision that nullifies Descartes' dualism, eliminates the idea of a separate soul, and places the cosmos at the center of spirituality.

Three centuries later, Einstein would recognize Spinoza as a master. He wrote to a rabbi in New York: "*I believe in Spinoza's God, who reveals himself in the harmony of what exists*." In that "God" there is no personal intervention, but order and beauty. In this sense, Deus sive Natura is very close to contemporary scientific sensibility, which sees the universe as a self-sufficient system governed by universal laws.

The appeal of this formula is that it does not reduce Nature to a cold machine. On the contrary, it imbues it with immanent sacredness. For panpsychism, this vision is fertile ground: if everything is a manifestation of a single substance, it is plausible that every fragment of the universe also possesses an internal aspect, an elementary "feeling."

In Spinoza's court, instead of angels and miracles, we find geometry, ethics, rational understanding, and respect for every being. In a letter from 1675, he confesses that he gave up publishing certain ideas for fear of religious hostility. Yet he knew that the seed had been planted: Deus sive Natura would survive well beyond his era.

Today, defining this formula means not only understanding Spinoza's philosophical position, but also recognizing its relevance. It is an invitation to think of the universe as an inseparable intertwining of matter and spirit, a field in which science and philosophy can meet without rigid boundaries.

What is panpsychism?

Panpsychism is an idea as ancient as it is disturbing: everything that exists possesses, in some form, a mental dimension. We are not talking about articulated consciousness, such as that of humans. We are talking about a primordial 'feeling', an inner aspect that accompanies matter.

The pre-Socratic philosophers had already outlined this hypothesis. Thales (in the 6th century BC) argued that "everything is full of gods," meaning that life or the soul permeates the entire cosmos. Plato, in Timaeus, describes a world generated by a anima mundi, capable of binding every element of the universe into a living fabric. Later, Giordano Bruno, during the Renaissance, spoke of an infinite multiplicity of worlds, each traversed by a vital principle.

Panpsychism is clearly distinct from other theories of the mind. Cartesian dualism separates mind and matter into two different substances: *res cogitans* and *res extensa*. Reductionist materialism argues that consciousness arises only from complex physical processes, such as those of the human brain. Panpsychism, on the other hand, does not separate or reduce: it attributes the same fundamental status to consciousness as it does to matter. Like mass or energy, consciousness is present in every fragment of reality.

This idea has influenced some unexpected thinkers. In the 17th century, Leibniz imagined the world as composed of *monads*, small indivisible entities, each endowed with perception. At the end of the 19th century, William James considered panpsychism a possible way to explain the continuity between life and consciousness. In the 20th century, Alfred North Whitehead reformulated it in terms of a "philosophy of process": every event has a physical and a mental aspect.

In contemporary science, panpsychism is back in circulation. Not as a poetic belief, but as a theoretical hypothesis. Some neuroscientists, such as Giulio Tononi, propose theories that see consciousness as a diffuse property, graded in intensity. In physics, some quantum interpretations suggest that the 'collapse' of wave functions may involve a role for consciousness or information as constituent elements.

Literary and cinematic culture has also taken up the theme. Arthur Koestler and Ursula K. Le Guin have described universes in which matter 'breathes' thought. In Terrence Malick's *'The Tree of* Life', the narrative mixes cosmology and introspection, evoking a living universe.

Panpsychism is fascinating because it shifts the focus of reflection. It asks not only 'what is the mind?', but 'how extensive is the mind?'. It does not limit itself to studying consciousness as a human phenomenon. It imagines an inner cosmos, where a stone, a tree, a star contribute, in different ways, to an enormous symphony of perception.

If we accept this premise, the relationship between man and nature changes. We are not surrounded by inert matter, but by realities that share, in an elementary form, something of what we call "experience." It is a thought that may seem radical, but one that over time has spoken to philosophers and scientists alike, and today dialogues with the most pressing questions of consciousness and life.

Deus sive natura and panpsychism: two horizons of universal thought

In the 17th century, Baruch Spinoza gave philosophy a formula that was destined to endure: Deus sive natura. This expression unites God and Nature in a single infinite, eternal substance, the cause of itself and of everything. For Spinoza, God is not a distant sovereign, but the very reality in which we live, made up of immutable laws.

His Amsterdam was rife with religious and political tensions. Spinoza, expelled from the Jewish community for his radical ideas, developed a vision in which every element of the universe was part of an absolute order. Deus sive natura is an ontological manifesto: everything is God, and God is everything. There are no boundaries between creator and creation.

This perspective clearly belongs to pantheism. It interprets every aspect of the world as a direct manifestation of the divine. It does not assign individual "consciousness" to everything: rather, it sees the totality as a single perfect being.

Panpsychism starts from another question: is consciousness a rare phenomenon or is it a universal property of matter? According to this theory, every element of the world—a man, a dog, a tree, a rock, and even an electron—possesses a form of inner experience. This experience may be infinitesimal, barely a rudimentary "feeling," but it is real.

The pre-Socratic philosophers had already intuited this idea. Anaximenes (6th century BC) identified air as the vital and thinking principle. Plato, in Timaeus, spoke of a 'world soul' that permeated everything. The Stoics saw the cosmos as a living being, whose 'reason' – the logos – was present even in what appears inert.

In the 17th century, Leibniz formulated the theory of monads: indivisible units endowed with perception, even when they show no apparent activity. In the 20th century, William James defined panpsychism as 'one of the oldest and most persistent hypotheses' for explaining the relationship between matter and mind.

The difference between 'Deus sive natura' and panpsychism is clear.

In Spinozian pantheism, the world is God. There is no distinction between matter, thought, and divinity: it is an ontological monolith. Consciousness is the prerogative of the whole, not of individual elements as autonomous entities.

In panpsychism, each entity has its own form of consciousness. It is not necessarily divine, nor is it necessarily united with all others in a supreme being. It is a mosaic of r small minds, from the elementary level of particles to the complexity of the human brain.

If we look at a forest, the pantheist sees an image of the divine in the whole. The panpsychist sees hundreds, thousands of individual consciousnesses: trees, plants, insects, molecules, each with its own way of "feeling."

Many cultures have shown panpsychist insights. Japanese Shinto recognizes a *kami* in every stone or mountain. In Mahayana Buddhism, universal consciousness permeates every phenomenon without being identified with a personal God.

Today, the discussion extends to the frontier of quantum physics. Roger Penrose and Stuart Hameroff have hypothesized that quantum phenomena in neuronal microtubules may underlie consciousness. Some scholars suggest that if consciousness were rooted in universal quantum processes, even the simplest objects could possess some form of experience. This would not be a return to pantheism, but an unprecedented scientific confirmation of an ancient idea of panpsychism.

"Deus sive natura" and panpsychism arise from the same wonder at the cosmos, but they follow different paths. Pantheism seeks divine unity, while panpsychism distributes the light of consciousness in every grain of reality.

Spinoza, writing that "all that is, is in God," proposed a vision that broke down the boundary between creator and creation. Panpsychism, as old as human thought, asks us instead to imagine a universe teeming with silent minds. Two paths to the same mystery: understanding how being and feeling are intertwined in the fabric of the world.

Attraction to the living cosmos...

Panpsychism exerts a fascination that goes beyond philosophy. It is a vision that invites us to look at the night sky, a tree, or even a stone with an extra thought: what if everything had an "inside," a form of primordial experience?

This attraction stems from an ancient impulse. Thales, at the mouth of the Aegean Sea, saw a widespread vital principle in the flow of rivers and the wind. Plato placed a soul in the world itself, linking the order of the cosmos to its invisible presence. Giordano Bruno, in 1584, wrote that *'nothing is dead in* infinity', imagining a universe dotted with living worlds.

The poetic vision of the living cosmos has influenced literature. In the 19th century, Walt Whitman sang of the "molecular brotherhood" that unites the human body with the earth and the stars. In the 20th century, Ursula K. Le Guin constructed entire planets where matter perceives the intentions of those who inhabit it. Each narrative opens a window onto the idea that the mind does not belong only to humans.

Many scientists admit to an intellectual attraction to this hypothesis, even if they treat it with caution. After discovering the electron equation, physicist Paul Dirac spoke of the "beauty" of nature as evidence of a profound order. Some neuroscientists, such as Christof Koch, wonder whether consciousness is widespread in embryonic forms even outside complex nervous systems.

From an existential point of view, panpsychism changes the way we live. If everything participates in a continuum of perception, then the relationship between humans and the environment becomes reciprocal. The forest listens as much as we listen to the rustling of leaves. The sea maintains a rhythm that can resonate with the beating of the heart.

Many find comfort in this. We are never completely alone if the universe is woven from a diffuse consciousness. The idea that a distant star, a speck of dust, or a gravitational wave 'participates' in the same principle that moves our thoughts creates a sense of belonging.

The appeal of the living cosmos is also aesthetic. NASA photographs showing spiral galaxies or iridescent nebulae become more than just scientific images. For those who accept the possibility of a diffused consciousness, they become portraits of an immense organism made of light, matter, and thought.

Panpsychism, in this sense, is an invitation to see the world as a companion rather than an object. It is a vision that combines scientific curiosity, poetic intuition, and the desire to feel part of a whole. It is not just a philosophical concept: it is a perspective capable of transforming everyday experience into cosmic participation.

Panpsychism and pantheism. Two ideas that seem similar, but are not.

The motto 'Deus sive natura', coined by Spinoza in the 17th century, united God and Nature in a single substance. With this formula, the Dutch philosopher questioned the idea of a personal and transcendent God, describing him instead as the infinite set of natural laws and phenomena. This vision is the quintessence of pantheism: everything is God, God is everything.

But panpsychism is another story. It does not say that everything is God. It says that everything has a form of consciousness. Panpsychism is an ancient philosophical hypothesis that attributes to every entity in the universe a form of inner experience—perhaps infinitesimal—but still real. It is not theology, it is metaphysics of the mind.

Anaximenes, a pre-Socratic philosopher of the 6th century BC, already saw air as a vital and conscious principle. Plato, in Timaeus, suggested that the soul of the world permeated every part of the cosmos. However, this 'cosmic soul' did not coincide with an omnipotent God, but with a network of life and widespread perception. Herein lies the first difference with pantheism: panpsychism does not elevate the whole to divinity, but recognises it as an immense fabric of consciousness.

The Stoic philosopher Chrysippus argued that even a stone, though devoid of thought, possessed a 'spark' of logos, an internal logic. In his school of thought, the universe was a living organism, but not a single God. In contrast, Giordano Bruno's pantheism in the 16th century saw nature itself as a direct manifestation of the divine and did not dwell on the possibility that everything had its own individual consciousness.

An example clarifies the distinction well. For the pantheist, a forest is God in one of his infinite forms. For the panpsychist, every single tree—and every cell of the tree—possesses a form of elementary perception, independent of the idea of God.

In ancient panpsychism, consciousness was linked to the 'psyche', the breath of life. In modern times, thinkers such as Leibniz described the world as a collection of monads, each endowed with perception. No monad is a dead fragment of matter; all participate in an inner life, even if unconscious.

Today, some theories link panpsychism to quantum physics. Sir Roger Penrose and Stuart Hameroff have hypothesized that microtubules in neuronal cells may host quantum phenomena linked to consciousness. Some more daring interpreters suggest that if consciousness emerges from quantum structures, then even elementary particles may possess a rudimentary form of experience. It is an unexpected return to a view close to panpsychism, but with the language of contemporary science.

In Eastern traditions, the panpsychist concept is widespread in other forms. Japanese Shinto attributes a kami, or spirit, to every mountain, river, and stone. Mahayana Buddhism speaks of a "universal consciousness" that permeates all phenomena, without defining it as God. This shows how pantheism and panpsychism have often intersected, but remain conceptually separate.

In the 19th century, William James recognized that panpsychism was "one of the oldest and most persistent hypotheses in the philosophy of mind." James, a pragmatist, saw this theory as an elegant way to un te mind and world, without resorting to dualisms or external deities.

Was Spinoza a pantheist or a panpsychist?

Spinoza is considered a pantheist, not a panpsychist. His famous motto Deus sive natura — "God, that is, Nature" — clearly expresses the idea that God and the world are the same infinite entity. For Spinoza, all reality is a single eternal substance, endowed with attributes such as thought and extension, but not divided into autonomous consciousnesses. Consciousness, in his view, does not belong to every single fragment of matter, but is a manifestation of that single universal substance.

Panpsychism, on the other hand, asserts that every entity, even the simplest, possesses a form of elementary experience. Here, consciousness is distributed, molecule by molecule, and not concentrated in the unity of the whole.

In summary, in Spinozian pantheism, the whole world is God.

In panpsychism, everything has its own consciousness, but not necessarily divine.

However, Spinoza, while being firmly a pantheist, shares some suggestions with ancient panpsychism in the way he conceives the deep connection between mind and matter.

In pre-Socratic philosophers and some later traditions, panpsychism arises from the idea that every element of the universe possesses some form, however minimal, of perception or vitality. Anaximenes attributed to air not only a material function, but also that of a vital and rational principle. Plato, in the Timaeus, spoke of the 'soul of the world', an intelligence that envelops and enlivens everything. The Stoics saw in the logos a cosmic reason spread everywhere.

Spinoza, in a different way, asserts that the only substance in the universe has two main attributes: thought and extension. Everything that happens on the material plane has a counterpart on the plane of thought. This parallelism is reminiscent of panpsychist sensibility: the recognition that what we call 'matter' is never devoid of a mental aspect.

However, the similarity ends at the point where panpsychism fragments experience into a multitude of small consciousnesses. Spinoza does not speak of individual minds nested in every particle; for him, there is only one infinite mind, of which every particular thought is a part. It is an echo of the Platonic 'world soul', but inextricably linked to a strictly monistic conception: everything is God.

In short, Spinoza shares with ancient panpsychism the idea that matter and thought are inseparable, and the vision of a cosmos that is intrinsically 'mentalized' or endowed with a thinking principle.

But he diverges because he attributes that thought to a single total reality, not to a myriad of autonomous consciousnesses.

The development of pantheism and panpsychism.

The formula "Deus sive Natura" belongs to Baruch Spinoza. Spinoza wrote it *in Ethica*, a work that sought to build bridges between God and the physical universe. His thesis was radical: God is not a being separate from the world, but coincides with the totality of nature, without exception. There is no metaphysical 'elsewhere': the whole, in its immense variety, is divinity itself.

This concept, which can be defined as pantheism, has very ancient roots. Already in the Vedic texts of India, there are verses that describe the cosmos as the body of a universal divinity. In the 6th century BC, Heraclitus stated that the Logos permeates all phenomena, like a diffuse intelligence. And in Pythagorean philosophies, there was talk of a universal harmony that connects heaven and earth through numbers, as if it were an invisible mind.

The step from pantheism to panpsychism is short but decisive.

Pantheism sees God and the world as identical. Panpsychism adds: if God coincides with nature, every part of nature must possess some form of consciousness, however minimal. In Greece, Anaxagoras introduced the concept of Nous, a cosmic mind that orders matter. In the Islamic Middle Ages, philosophers such as Al-Kindi and Avicenna discussed how the soul could extend beyond the human body, hypothesizing a mental presence in natural phenomena.

In the European Renaissance, Giordano Bruno took up these themes. His works describe an infinite universe, populated by infinite worlds, each animated by a spark of life and consciousness. Bruno was condemned for heresy, but his idea profoundly influenced the nascent scientific culture, fueling the vision of a living cosmos.

In the 19th century, panpsychism found new avenues. Gustav Fechner, a German physicist and philosopher, declared that the Earth is a living being with its own mind, like an organism. William James, the father of American pragmatic psychology, wrote that there is no clear line between 'inanimate' matter and conscious mind: consciousness could be a universal attribute, varying in intensity but not in presence.

At the same time, the scientific debate on electricity, magnetic fields, and biological life fueled hypotheses about the 'sensitivity' of matter. Many physicists and philosophers wondered whether energy itself could be a primitive form of consciousness.

Oxford, Cambridge, Harvard: these are universities where, between the late 19th and early 20th centuries, panpsychism was discussed in small seminars. In Germany, in Jena, the Romantic circle saw Goethe as the spokesperson for a creative and intelligent nature. In Italy, thinkers such as Bertrando Spaventa and Antonio Labriola grappled with translations of Spinoza and Bruno, seeking to reconcile philosophy with evolutionary science.

The 20th century, marked by Einstein and quantum mechanics, opened up new scenarios. If matter is composed of entities in constant relation, if particles can be ' ly' correlated at a distance, then perhaps consciousness could emerge from this network of connections. Some physicists and philosophers, such as David Bohm, introduced the idea of an "implicate order" of the universe, similar to a diffuse mind that manifests itself in observable phenomena.

Today, panpsychism has returned to philosophical and neuroscientific discussion thanks to figures such as Philip Goff, who argues for the plausibility of a universal consciousness as an explanation for the "hard problem" of the mind: how matter becomes subjective experience. Although not everyone shares this idea, the question remains open.

What remains of pantheism today?

Pantheism, understood in its classical formulation—God as identical to the totality of nature—has not disappeared, but today it has taken on different forms than in past centuries.

It is no longer dominant in philosophical debate, as it was in the days of Spinoza or Bruno, but it continues to influence thinkers, scientists, and cultural movements, often under new labels.

Today, 'pantheism' in the strict sense is hardly ever mentioned in universities or scientific journals. Concepts such as 'neutral physicalism', 'panpsychism' or 'spiritual naturalism' are preferred, as they can be compatible with certain pantheistic premises without using theological language.

The meaning remains similar: seeing nature as a self-sufficient and sacred reality, recognizing a profound unity between all things. In the field of environmentalism, for example, many ethical and philosophical currents recover an implicit pantheism, treating the Earth as a living organism worthy of intrinsic respect.

Furthermore, contemporary physics and cosmology offer images of the cosmos that are close to the pantheistic view, even if expressed in secular terms: a universe without an "outside," structured by universal laws, where energy and matter are manifestations of the same reality.

Even in the religious sphere, there are churches and movements—such as certain circles of neo-paganism, Western Buddhism, and non-dogmatic Christianity—that reinterpret God as cosmic totality, effectively reviving the pantheistic core, albeit in forms more compatible with contemporary sensibilities.

Pantheism today. From philosophical tradition to new visions of the cosmos.

Pantheism, in its classical forms, seems to belong to an era of great philosophical systems and all-encompassing metaphysical visions. However, today it presents itself in new, secular, and scientific guises. Its essence remains: the belief that the divine, or what can be called the "universal principle," coincides with the totality of nature.

The difference lies in the language used to express it. In past centuries, it was theological or philosophical language; today, it is often the language of physics, ecology, and environmental psychology.

Contemporary physics, particularly cosmology, describes a universe that has no boundaries 'outside' itself. The laws governing matter and energy are universal and interconnected. Some scientists interpret this order as a form of cosmic unity which, while avoiding religious terms, recalls Spinoza's vision.

Physicist John Archibald Wheeler, for example, spoke of a 'participatory' universe, where observer and cosmos mutually influence each other's state. Without metaphysical proclamations, but with an intonation reminiscent of the idea of a thinking nature.

Many environmental movements incorporate an implicitly pantheistic way of thinking. Arne Næss's deep ecology considers the Earth as a living entity, worthy of intrinsic respect, regardless of the use that man may make of it. The ethical charter of the Gaia movement, also inspired by the theories of James Lovelock, echoes ancient words: the Earth is an organism, and we are its cells. It is a 'science' that touches spiritual chords.

In contemporary Christianity, there are groups that reinterpret God as a cosmic totality, eliminating the separation between Creator and creation. These groups spread a vision in which the sacred is not in a temple, but in the forest, in the galaxies, in the water of a river. It is a living pantheism, even if often not declared as such.

Theories on quantum connections, from the work of David Bohm to the current thinking of Carlo Rovelli, open up avenues in which physical order and universal interconnectivity take on an almost metaphysical aura. Some philosophers, such as Philip Goff, return to panpsychism to say that consciousness could be a fundamental aspect of reality, just like space and time. In this framework, pantheism becomes compatible with a scientific view of mind and matter.

Renewed pantheistic ideas circulate at philosophy festivals, such as 'Filosofia' in Modena, Carpi, and Sassuolo, or at conferences on consciousness held in London, New York, and Tokyo. Scientific conferences and digital cultural movements, such as those popular in online communities on the theme of 'the cosmos as an organism', are places of discussion where pantheism appears without explicitly declaring itself.

Is it justified to refer to Spinoza as a precursor of panpsychism?

Today, there is some confusion between pantheism and panpsychism, especially in non-specialist thinking.

The overlap arises from the fact that both share a central idea: the universe is not merely a sum of inert objects, but a totality animated by a principle that is, in some way, 'mental' or 'spiritual'.

The two terms are confused for several reasons. The first is the use of non-technical language. Outside the academic sphere, many people use 'pantheism' to mean generically that 'everything is divine' or 'everything is alive', without distinguishing whether they are talking about God as nature (pantheism) or consciousness distributed throughout nature (panpsychism).

The influence of spiritual and ecological issues must be considered. In ecological movements, New Age, and some contemporary philosophical currents, the ideas of a sacred cosmos and a universal mind are intertwined. The boundary between giving nature a divine character and attributing consciousness to it is not always perceived.

Some confusion also stems from their common historical origin. Many of the roots of panpsychism can be found in pantheistic texts and authors; as a result, in popular discourse, the concepts become intertwined and end up appearing as sides of the same coin.

In reality, the difference is conceptual. In pantheism: God and Nature are identical; nature is divinity itself, but not necessarily 'conscious' in the human sense.

In panpsychism: every fragment of reality, from galaxies to atoms, possesses some form of consciousness or experience, however minimal; this does not necessarily imply a God.

Pantheism is primarily a theological or metaphysical ontology; panpsychism is a theory of the mind extended to the entire cosmos.

Can Spinoza be considered a precursor of panpsychism? Indeed, it is justified to see him as a precursor, albeit with caution. Spinoza does not openly state that every single fragment of reality has consciousness, but his doctrine of the "two fundamental attributes" — thought and extension — implies that the mental dimension is coextensive with the physical dimension.

SPINOZIAN PARALLELISM.

In Spinoza's philosophy, Substance (which he identifies with God or Nature) has infinite attributes, but human beings are aware of only two: thought and extension. These are not two separate "parts" of reality, but two ways of perceiving and describing the same unique reality.

In other words, everything exists simultaneously in terms of extension (physical, bodily dimension) and in terms of thought (mental or logical dimension). This correspondence is perfect and necessary: every state of the body has a correlate in the "state of mind" and vice versa.

This doctrine is known as parallelism. There is no physical cause that directly produces a mental fact, nor is there a mental cause that directly produces a physical fact.

Rather, body and mind are two coextensive manifestations of the same reality (Substance).

It is precisely this coextensiveness that leads some contemporary interpreters to see Spinoza as a precursor of panpsychism: if everything that exists also has a "mental side" in addition to a physical one, it seems that consciousness—or a form of experience—is widespread everywhere, at least in potential.

If God coincides with nature, and nature also possesses the attribute of thought, then the entire natural system possesses a mental aspect.

Many modern philosophers, from Fechner to Goff, have read this approach as a foundation compatible with panpsychism, even though Spinoza remained focused on ontology rather than a theory of consciousness.

Below, I propose a historical and conceptual synoptic table that clearly shows the evolution of pantheism and panpsychism, their points of contact and their divergences.

Table1—Comparison between pantheism and panpsychism.

Aspect	Pantheism	Panpsychism
Definition	Identifies God with Nature: everything that exists is a manifestation of the divine. Nature is sacred, but not necessarily "conscious" in the human sense.	Every element of reality possesses some form of experience or consciousness, however minimal. It does not necessarily imply a God.
Historical origin	Ancient roots: Stoicism, Neoplatonism, Renaissance philosophy (Giordano Bruno). Systematic formulation in the modern age with Spinoza.	Embryonic ideas in ancient thinkers (Anaximenes, Plotinus). Explicit development with Gustav Fechner (19th century) and contemporary discussions in the philosophy of mind.
Central figure	Spinoza: doctrine of the two attributes (thought and extension) → mind and body as coextensive dimensions of the same substance.	Fechner (soul of the Earth), William James (diffuse experience), today Philip Goff and other philosophers of mind.
Vision of the Cosmos	Divine cosmic unity: the world is God and God is the world; every being is part of the sacred whole.	Animated universe: every fragment, from subatomic matter to galaxies, has a mental dimension.
Focus	Ontology and theology: definition of being as identity between God and nature.	Philosophy of mind: nature and consciousness are widespread and fundamental aspects of reality.
Points of contact	Both affirm that reality has a universal dimension that is	They share the idea that the mind or spiritual principle is not confined to humans.

Aspect	Pantheism	Panpsychism
	inseparable from the whole of things.	
Key differen ces	Places the divine at the center; does not require that everything has consciousness, but that everything is a manifestation of God.	Does not necessarily speak of God; sees consciousness (or proto-consciousness) as a universal property of matter.
Releva nce	Revived in eco-spiritual forms, open theologies, scientific philosophical movements that see nature as a sacred whole.	Discussions in the debate on consciousness, neuroscience, quantum interpretations; basis for some ecological and cosmological views.

From the universal spirit to neuroscience.

Every era has sought a bridge between what we perceive as 'spirit' and what we measure as 'matter'. Panpsychism offers a surprising route for this journey. It starts with the hypothesis that everything, from the oak tree to the quasar, possesses an inner principle. This widespread feeling, which for Spinoza is intertwined with the concept of *Deus sive Natura*, finds a new language today in neuroscience and theoretical physics.

The ancients saw the cosmos as an organism. Plato's *"anima mundi,"* Leibniz's *monad*, Bruno's vital "breath" were not isolated figures. They were attempts to formulate a single idea: life is not an island, but an invisible continent that envelops everything. From this arises the common thread: the continuity between mind and world.

In the 20th century, biology and physics quietly rediscovered this link. Contemporary neuroscience takes this reflection even further. Giulio Tononi, with his Integrated Information Theory (IIT), treats consciousness as a fundamental, measurable property present in systems of varying complexity. Christof Koch, studying the brain and perception, admits to having found in panpsychism fertile ground for explaining forms of consciousness not limited to humans.

Quantum physics also provides insights. Roger Penrose hypothesizes that neuronal microtubules may host quantum processes related to conscious experience. It is a controversial theory, but it fits into the same historical trajectory: matter is not mute, but possesses a substrate of possibilities that human thought can intercept.

The path from the 'universal spirit' of the ancient philosophers to today's neuroscientific metrics is neither linear nor peaceful. It is a tapestry of insights and tools, poetry and laboratories. Yet it retains an unchanged core: the idea that mind and universe are two sides of the same substance.

This continuity can be described as a dialogue across the centuries. Between the meditative silence of Rijnsburg, where Spinoza ground lenses to see far away, and the control rooms of CERN, where scientists try to read elementary particles, there is the same tension: understanding whether reality is just a set of facts, or whether each fact carries an inner aspect.

From the conscious cosmos to everyday life

Thinking of the world as having an "interior" changes our relationship with the simplest objects. A tree is no longer just wood and leaves, but a silent interlocutor. An old stone in the garden becomes a witness to a shared history. Looking at the sky at night ceases to be an act of passive contemplation: it becomes a dialogue.

In everyday practices, this perspective produces concrete actions. Those who adopt a panpsychist point of view tend to take greater care of the spaces they inhabit, as if they were part of communal life. A house is not just a container, but a sensitive fabric. Even the way we consume or produce energy changes: we feel a debt of respect towards matter, because we do not consider it inert but, at least in part, part of the whole.

Contemporary arts play a role in this process of practical transformation. A film such as *"The Tree of Life,"* which places the story of a family in the cosmic vastness, can prompt the viewer to see their own life as a fragment of an immense organism. An installation such as "*The Weather Project*," after being visited, can make one more aware of light, air, and climate.

Ambient music is another concrete example. This is a musical genre that emerged in the second half of the 20th century, characterized by soft soundscapes, spatiality, and the absence (or near absence) of traditional rhythmic elements. Those who listen to certain compositions by Brian Eno report changing the rhythm of their daily activities. The sounds create a sense of connection with the environment, inducing an attitude of calm and attention.

The most profound implication is the redefinition of 'responsibility'. If everything has a form of participation in being, our actions become an integral part of it. The way we walk along a path or place an object down does not only concern us: it also concerns the context which, according to this view, perceives and reacts.

In short, panpsychism translated into art becomes practice when it shapes the way we see and act. Theory becomes habit, aesthetics becomes ethics, and the universe ceases to be a backdrop and becomes a co-protagonist of our existence.

Considering the world as alive and conscious is not just a speculative exercise. It is a way of experiencing reality. Spinoza formulated it in words that are destined to remain: *Deus sive Natura*. God and Nature are one and the same substance, infinite and indivisible. Every part of the cosmos, from a speck of dust to the stars, is an expression of this single being.

When we adopt this perspective, the practical implications also change. Caring for the environment is no longer a moral gesture towards something external, but an act of self-preservation: protecting the world means protecting what we are an integral part of. Everyday relationships take on a new intensity, because every encounter is a contact between manifestations of that same universal substance.

Modern science, observing the complexity of consciousness, seems to be approaching this philosophical framework. Neuroscience studies consciousness as a widespread property and, in part, intrinsic to different systems. Quantum physics, with its most daring hypotheses , suggests that the observer and the observed are linked by a creative relationship. Art, too, in giving form to a cosmos perceived as alive, translates this intuition into sensory and emotional experience.

In this intertwining, panpsychism reveals its greatest strength: bringing together seemingly separate fields. Philosophy, science, and art become three modes of investigating the same mystery. And everyday life becomes the place where these modes meet, transforming an ancient thought into present practice.

II. The origins of panpsychism.

To speak of panpsychism is to approach one of the oldest and most persistent ideas in human history: the perception that everything—from stones to the sun, from the wind to the tiniest creature—possesses an inner life, a form of consciousness. This intuition, which Spinoza would formulate centuries later as Deus sive Natura, originated in the depths of myth and developed in philosophy long before the modern era.

Its earliest roots lie in archaic myths and cosmologies. In the animist societies of Africa, Asia, and the Americas, the world appears to be populated by spirits and vital forces. In Polynesia, Tangaroa is the god of the ocean, but the ocean itself is a living creature. In ancient Egypt, the Nile is not just any river: it is Hapi, a deity who nourishes the Earth. Indian pantheism, expressed in the Vedas (circa 1500-500 BC), also identifies the universe with Brahman, energy and consciousness at the same time.

With the pre-Socratics, thought became structured. Thales of Miletus, born around 624 BC, stated that "*everything is full of gods*." This is not a metaphor, but the belief that everything hosts a vital spark. Anaximander saw matter as coming from the *ápeiron*, an infinite and indeterminate principle, similar to a cosmic womb that generates life and movement.

Platonism brings this vision to an orderly and rational plane. For Plato, the cosmos is a living organism created by the Demiurge and animated by a World Soul. In Timaeus, written around 360 BC, the philosopher describes the universe as a perfect body, capable of perceiving and thinking, where geometry and life are inseparable.

With Aristotle, the concept of 'vital form' is refined. Born in Stagira in 384 BC, Aristotle introduces *entelechy*: the internal principle that guides every being towards its full development. A plant grows and flowers because it has an intrinsic purpose, just like an animal or a human being. This force is not separate from matter, but constitutes its most authentic reality.

Finally, the Stoics interpreted the cosmos as a single body, permeated by pneuma, the warm, vital breath that guarantees cohesion and rationality. Zeno of Citium, founder of the school in the 3rd century BC, taught that living according to nature means living according to reason, precisely because the whole of nature is active reason. Rome inherited this vision: Marcus Aurelius, in his Meditations, wrote that "everything that happens is right, because it is the result of a universal mind."

These stages—myth, the Presocratics, Platonism, Aristotelianism, and Stoicism—form a thread of thought that spans the centuries. A thread that unites the tribal village that speaks to the river, the philosopher who speculates on the Soul of the World, and those who today discuss quantum consciousness. That is why, when we talk about panpsychism, it is not just a theory: it tells the long story of a universal human intuition.

Myths and archaic cosmologies.

Long before philosophers wrote treatises, humanity viewed the world as alive. Every natural element was not an object, but a subject endowed with intention and power. This belief translated into rituals, stories, and images that we now define with terms such as animism, pantheism, and polytheism.

In animist thinking, already present in Paleolithic societies, the mountain possessed a spirit, the river had a voice, the animal acted as a brother or adversary. In Australian Aboriginal communities, the "Dreamtime" describes creation as a continuous act of song and transformation: rocks and waterways are ancestors, guardians of memory. Around the first millennium BC, Siberian shamans spoke to the "genie" of the forest before hunting. In these cultures, living meant negotiating with a network of invisible presences that were perceived as real.

Ancient pantheism took this idea even further: the divine was not only in everything, but coincided with everything that exists. In the Vedas of India, composed between 1500 and 500 BC, Brahman is not one god among others, but the universal being: energy, consciousness, and matter fused into a single substance. In China, Taoist thought, as early as the 6th century BC, spoke of the Tao as the original principle that pervades all reality, living or 'inanimate'. This conception did not require a specific temple: the temple was the entire cosmos.

Polytheism, on the other hand, fragmented the divine into a constellation of figures, but maintained the same vital basis. In archaic Greece, Zeus was not just a celestial god: the sky itself was a living entity that thought and acted. Poseidon did not inhabit the ocean—he was the ocean. In ancient Egypt, the sun was Ra, and each day was perceived as the actual journey of this deity across the sky. The Incas, in their capital Cusco, worshipped Inti, the sun, and Pachamama, the mother earth, without drawing rigid boundaries between matter and spirit.

These views had a direct impact on practical life. Sowing a field became a sacred act, cutting down a tree required ritual permission. This was not random superstition, but a cultural technology that taught respect and responsibility towards all forms of existence.

Panpsychism, in its most spontaneous form, was born here: in the belief that there was no clear separation between subject and object, between human and non-human. Everything was a 'participant' in the same cosmic conversation. It is in these mythical and religious roots that we find the seed of the idea that Spinoza, millennia later, would formulate as Deus sive Natura.

ANIMISM. THE WORLD AS A NETWORK OF PRESENCES.

In Australian Aboriginal societies, the Dreamtime is not just a creation myth. It is a living map. Every hill, every watercourse is a trace left by ancestors, primordial beings who have transformed themselves into elements of the landscape. Walking through the territory means traversing a biography of the earth.

In Siberia, around the first millennium BC, shamans approached hunting as a dialogue. Before hunting an elk or a deer, they spoke to the "lord" of the forest, asking permission to interrupt a life cycle. Here, animism manifests itself as alliance and responsibility: every human gesture is part of a cosmic reciprocity.

Among the natives of the Amazon, even today, the river is not just water. It is a 'grandfather', a being that listens and can nourish or punish. Fishermen do not cast their nets lightly, because the river has memory and will.

PANTHEISM. THE UNITY OF THE DIVINE AND THE UNIVERSE.

In the Indian Vedas (1500-500 BC), Brahman is the totality. He is not one god among others, but the universal essence, the energy that sustains matter and spirit together. Vedic chanting does not aim to please a separate deity, but to synchronize human beings with cosmic vibration.

In 6th-century BC China, Laozi describes the Tao as 'the way' and 'the mother of the world'. The Tao does not rule from above, but flows in everything: in the sprouting of a flower, in the ebb and flow of the tides, in the changing of the seasons. The Taoist temple has no walls, because worship takes place wherever the natural order reveals itself.

POLYTHEISM. A CONSTELLATION OF LIVING POWERS.

In ancient Greece, Zeus is not only the king of the gods. He is the sky itself, with storms and clear skies as forms of his action. Poseidon is the depth and fury of the ocean; the waves are blows of his will. Natural experience is inseparable from the divine presence.

In Egypt, Ra is the sun, but also the traveler who crosses the sky every day, fighting the forces of chaos in the night. The solar cycle is a lived narrative that marks human and cosmic time.

Among the Incas, Inti (the sun) and Pachamama (mother earth) are powers with which one lives in alliance. Sowing is an act of devotion; the harvest is a mutual gift between man and divinity. There is no clear separation between the physical and the spiritual, because the divine is matter and matter is divine.

These manifestations, different but converging, outline a cultural thread: every civilization has found a way to affirm that the cosmos is a place inhabited by consciousness. Whether it is an ancestor, a universal principle, or a plurality of gods, the idea that nature is alive spans eras and geographies.

Shamans and priests.

In many ancient cultures, there are figures who perfectly embody the panpsychist view of the world, acting as intermediaries between the visible and invisible dimensions, between humanity and the widespread vitality of nature. Their roles and practices are closely linked to the belief that everything—stones, trees, rivers, animals—has its own consciousness or spirit.

In the cultures of the taiga and tundra, the shaman is the guardian of dialogue with the spirits of the forest and the earth. Before a hunt, he enters a trance state through drumming and rhythmic dancing; in this state, he 'travels' to the spiritual world to ask permission to kill an animal, offering respect and rituals of thanksgiving in return. These gestures are not mere symbolism, but concrete acts to maintain the balance between man and the environment, in line with a panpsychist way of thinking that sees life as a network of cosmic alliances.

Among the communities linked to the Dreamtime, the spiritual curator has the task of keeping the connection with the ancestors alive. He walks in sacred places and 'reactivates' stories by singing the routes traveled by primordial beings. Each song is an act of creation that reconfirms the intrinsic consciousness of the earth; without these rituals, there is a fear that the landscape may 'sleep' and lose its vitality. Here, the priest's function is also to preserve the living map of the territory.

In Vedic rituals (1500-500 BC), the priest is the intermediary between Brahman, the universal essence, and human reality. He recites hymns and offers ritual fire, considered a bridge between the earthly world and the divine totality. Every gesture—from pouring clarified butter as an offering to the rhythm of the formulas—aims to synchronize human action with cosmic vibration, reducing the distance between matter and spirit.

In ancient China, the Taoist master often lives in seclusion, observing the changing seasons and teaching that the Tao flows in everything. His practices include meditation, breathing exercises, and harmonization with natural cycles. He does not invoke a separate god, but attunes himself to the universal principle that permeates the world, expressing a form of practical pantheism.

In the temple of Ra or Amun, the priest celebrates the rising of the sun as a true divine journey. He lights fires, offers bread and beer, and recites formulas that 'nourish' the god so that he continues to sustain Maat, the order of the universe. Celebrating the deity also means supporting the natural cycle, because the god is the living manifestation of a physical phenomenon.

These figures, although different in culture and ritual style, share a fundamental assumption: material reality is animated, and human beings must interact with it through respect, communication, and mutual exchange. In this sense, they are the first concrete interpreters of a panpsychist vision, long before philosophy codified its theories.

III. Pantheistic suggestions in Eastern civilizations.

Let us imagine a remote time, before the need to separate science and religion, even before Greek philosophy could give a name to the intuitions of the human soul. In that ancient world, the universe was not seen as a collection of inert objects, but as a living fabric, pulsating with invisible presences.

Under the Egyptian desert sun, among the waters of the Nile, in the boundless Mesopotamian horizons, on the banks of the Ganges or in the silent valleys of China, ancient civilizations perceived a common principle: the divine did not "stand above" the world, **it was** the world. The stars, animals, rivers, mountains—everything was part of a cosmic order in which matter and spirit were inseparable.

These ideas, which we now recognize as pantheistic, anticipate the radical identity between God and Nature that Spinoza would express with his famous "Deus sive Natura." They also lay the foundations for a panpsychist view: the idea that every fragment of reality, from the largest galaxy to the smallest grain of sand, participates in some form of consciousness or vital principle.

This chapter will traverse sacred rivers, creative myths, celestial laws, and invisible paths to rediscover in the eyes of the ancients the awareness that the universe itself is the temple.

The sacred river and the divine order. Ancient Egypt .

In Egyptian civilization, the divine was not located outside the natural world. The divine permeated every aspect of reality, intertwining with the cosmic order of Ma'at, the principle of balance and justice that governed both human destinies and the course of the stars.

The Nile was the beating heart of this order. A river of life, it was revered as the god Hapi, bringer of the annual floods that nourished the earth. Water was not just an agricultural resource. Water was a concrete manifestation of universal harmony.

Temple inscriptions dedicated words to Amun-Ra that sound extraordinarily modern today: "*He who is in everything.*" The idea was clear. The god did not reign over nature, but acted within it, through the rising of the sun, the song of the ibises, the flooding of the river.

Every year, at the height of summer, priests and scribes measured the waters of the Nile in the nilometers of Elephantine or Memphis. This seemingly technical act was a sacred ritual. The measurement indicated whether the harvest would be abundant or poor. It was interpreted as a sign of divine will.

The pharaoh, considered the son of Ra, had the task of maintaining Ma'at. To rule meant to protect the balance between men, gods, and the river. An unjust reign was believed to be capable of disturbing the flooding of the Nile, breaking the harmonious rhythm of the universe.

Many scholars see this view as a primitive form of pantheism. The divine principle was not distinct from matter. Water, earth, and sky were part of a single living organism.

This perception, developed over three thousand years ago, has left its mark on contemporary ecological thinking. Some theologies of nature today propose a similar vision: the Earth as a sacred entity, where life and the divine coincide.

Thus, the Nile of ancient Egypt still speaks to us. Not only as a geographically real river, but as a universal symbol of an indissoluble link between the order of nature and the breath of the divine.

The pantheistic vision of the ancient Egyptians in temple inscriptions.

Many Egyptian temples contain inscriptions which, when read carefully, clearly reveal the pantheistic conception of a divine presence diffused throughout every aspect of the universe, and not confined to a separate 'afterlife'. Here are some significant examples:

In the Temple of Karnak, *the Hymn to Amun-Ra*, an inscription carved on the eastern wall of the temple dating back to around the 18th dynasty, reads:

"You are the sun that shines, the air that every nose breathes, the water that flows in the river, the seed that germinates in the fields."

Here, the deity is not described as a distant object, but as an essence present in natural phenomena, vital for humans, animals, and plants.

In the Hypostyle Hall of the Temple of Luxor, a series of reliefs reads:

"Amon, lord of the Two Lands, dwells in the tree that gives shade, in the flower that opens in the morning, in the wind that brings the flood of the Nile."

The fusion of plant images and atmospheric cycles clearly indicates that the deity operates from within nature and is its living substance.

In the Temple of Esna, the *Hymn to Khnum and Hapi*, a text from the Ptolemaic period, celebrates Hapi by saying:

"Hapi, belly of the Earth, he who feeds every city, you are the sap that flows and the life that is generated in water."

This biological language highlights the perception of the river as a divine organism, with a "body" that coincides with that of the earth itself.

The Great Hymn to Aten from the Amarna period engraved in the Tomb of Ay reads:

"You are in the mother who nourishes, in the newborn who breathes, in the flight of birds and in the swimming of fish."

Here, the sun god has no boundaries: he is present in animal life, human life, and nature in motion.

These texts reflect that matter and spirit are not separate entities, but two faces of the same universal reality.

Heaven and Earth in a single breath: Mesopotamia.

In the plains of Mesopotamia, where the Tigris and Euphrates rivers mark the rhythm of the seasons, the ancient Sumerians and Babylonians saw heaven and earth as a single living body. They were not two separate realities, but two expressions of the same divine principle. Heaven, An, and earth, Ki, formed an original couple, father and mother of all things.

Matter was not inert. Matter was the very body of the gods. The cosmos vibrated with its own life, and the natural elements were perceived as organs of a great sacred organism. Rain was the breath of An. The fertile earth was the womb of Ki.

This vision emerges clearly in myths. The "Enuma Elish," a Babylonian poem of creation, tells that the universe was born from the transformation of a cosmic being: the goddess Tiamat, primordial mother of salt waters. Killed by the god Marduk, Tiamat does not disappear. Marduk opens her body and with one half forms the sky, with the other the earth. Creation is an act of metamorphosis. The divine continues to live as the very structure of the world.

Tablet IV of the Enuma Elish reads:

"He divided her carcass like a fish, one half above, as the roof of the sky, the other below, as the foundation of the earth."

There is no god who stands outside his work. There is a god who remains within his creation, transforming himself into it. This image is extraordinarily close to pantheistic conceptions: the divine as the totality of the cosmos, inseparable from the matter that composes it.

In everyday life, this idea was celebrated in rituals and architecture. In Ur, the priests of the temple of Nanna, god of the moon, climbed onto the terraces of the ziggurats to observe the sky. They did not merely scrutinize the stars as natural phenomena. They interpreted their geometries as the very language of the gods.

The king, as guardian of the cosmic order, had a duty to maintain harmony between heaven and earth. This was not a symbolic task. The justice of the sovereign was believed to influence the waters of the rivers and the fertility of the soil. An unjust kingdom could break the common breath of An and Ki.

Today, this ancient perception resonates in ecological theologies and philosophical reflections on panpsychism. The notion that matter is the bearer of life and consciousness is not new. In Mesopotamia more than four thousand years ago, it was already rooted in the language of myths, ceremonies, and laws.

Heaven and earth, united in a single breath, continue to speak to us through clay tablets and ziggurat ruins. They remind us that seeing the world as a divine body means, first and foremost, recognizing that we are part of this same living organism.

Mesopotamian thought in modern interpretations of pantheism.

Modern interpretations of pantheism and deep ecology find one of their oldest archetypes in Mesopotamian thought.

In the Sumerian and Babylonian traditions, An (sky) and Ki (earth) were perceived as parts of a single living organism. Creation, as narrated in the Enuma Elish, was not an external act of a separate god, but an internal transformation: the body of Tiamat, the primordial goddess, literally became the sky above and the earth below. In this paradigm, the divine does not inhabit nature, the divine is nature.

This view corresponds perfectly to the core of modern pantheism, especially in Spinoza's formulation: "God or Nature" as the only universal substance, in which everything that exists is part of an inseparable continuum.

Deep ecology takes up this sensibility but reformulates it in scientific and ethical terms: the Earth is considered an integrated living system, almost an "organism" — an idea that echoes Mesopotamia, where the water cycle, the fertility of the fields, and the movement of the stars were perceived as vital functions of the cosmic body.

In both perspectives, ancient and modern:

- The separation between matter and spirit is denied.
- The natural order is sacred and binding, not only for religious reasons, but as a condition of existence.
- Human action must maintain the harmony of the "breath" between heaven and earth, or between the biosphere and the cosmos.

Today, ecological philosophers and theologians speak of Gaia as a living and conscious entity, a vision that overlaps with the Mesopotamian idea of a divine world in which every part, from the wind to the river, is an expression of universal life.

The Tao: the Way as the living structure of the world .

Ancient Chinese thought, in the Zhou era (11th–3rd century BC), conceived of the Tao as a universal law, invisible and formless, which generates everything and to which everything returns. It was not an anthropomorphic god. It had no face, nor individual will. The Tao was the very rhythm of the cosmos, a silent flow that nourishes, transforms, and dissolves.

Historical origins of the Tao: from ritual to universal law

In China during the Zhou dynasty, between the 11th and 3rd centuries BC, politics and spirituality formed an indivisible fabric. The sovereign reigned not only by dynastic right, but because he was invested with the Mandate of Heaven (Tian Ming). This Heaven was not a personal god, capable of commanding or punishing as in theistic religions. Heaven was an ordering principle, a force that permeated the world as a moral and cosmic field, immutable in its purity and rhythm.

The Mandate of Heaven established that power had to be exercised in harmony with the universal order. If the ruler betrayed this harmony, wars or famines were interpreted as signs of the loss of the Mandate. This conception transformed the task of government into an act of harmony with the laws of nature. It was not a matter of obeying an arbitrary divine will, but of maintaining the balanced flow of the world.

Parallel to the rituality codified by Confucius, a more contemplative way of thinking developed. In the first texts attributed to Laozi, collected in the *Tao Te* Ching (6th–4th century BC), the Tao was described as "*the invisible mother of all things.*" Laozi wrote: "*The Tao generates and nourishes, grows and completes, but does not dominate.*" These words convey the idea of a creative force that operates without the aim of control and without hierarchies.

The Tao is not born and does not die. It does not begin with a single act as in Mesopotamian myths, where the world arises from the body of a primordial deity. In Taoism, creation is a continuous and spontaneous process. Every season, every natural cycle, every event is a manifestation of this eternal flow.

A Taoist tale tells of a hermit who lived in the mountains of Luoyang. Observing the course of a stream, he explained to his disciples that the Tao flows like water, overcoming every obstacle with its persistence. When a boulder falls into the riverbed, the water does not oppose the obstacle, does not force the situation, does not make unnecessary efforts. It finds a new path. The wise man must do the same.

Many places in China, such as Mount Hua Shan, were considered centers of high concentration of the Tao. Monks would retreat there to meditate, convinced that the geography itself could favor the perception of the universal rhythm. Every rock, every tree, every breath of wind was seen as part of an indivisible cosmic organism.

This perspective fits into the great history of pantheistic and panpsychist thought. The Tao is a living universe in which everything has a form of 'participation' in the universal law. There is no division between the principle that orders and that which is ordered. Everything is in the same breath, in an unbroken chain of generation and return.

In modern times, scholars of deep ecology have recognized in the Tao a philosophy close to the vision of the Earth as an integrated organism. It is the same intuition that today drives physicists and philosophers to speak of universal interconnection, where parts are never isolated but connected by an invisible fabric. Just as the Tao described more than two thousand years ago.

The nature of the Tao: unity and harmony of opposites .

The Tao is a constant presence. It is a flow that does not distinguish between good and evil, victory and defeat, birth and death. In Taoist thought, every reality has its opposite, and existence arises precisely from the relationship between these polarities.

The principle of *yin and yang* expresses this eternal dynamic. Yin is shadow, stillness, the feminine, the rest of the night. Yang is light, motion, the masculine, the full energy of the day. Neither is superior to the other. Neither is complete without the other. The curve of the famous symbol (half black, half white, each with a seed of the opposite color) reminds us that in every light there is shadow and in every shadow there is a point of light.

In the 8th century, the painter Wu Daozi said he perceived the whole world as "a single breath" expanding between heaven and earth. According to him, the wind was not separate from the rock, and the river was not distinct from the clouds. Everything moved as a single body. Wu painted landscapes where water wrapped around mountains like veins around a heart. This vision, so tangible to him, is the Tao in its purest form: a living organism where each part supports and depends on the other.

The life of farmers along the Yellow River reflects the same philosophy. When the floods came, their land was enriched with fertile silt. They could plant wheat and millet with the certainty of abundant harvests. But the same floods could turn into a catastrophe, submerging fields and destroying homes. Joy and fear were two sides of the same event. For the farmer, the flood was a direct manifestation of the Tao, both generous and cruel, never just one or the other.

The Tao does not seek to eliminate the tension between opposites. On the contrary, it maintains it, regulates it, transforms it into movement. In Taoist philosophy, balance does not mean immobility. It means the orderly alternation of forces. The life of the fish, which swims upstream to lay its eggs and then lets itself be carried by the current back to the sea, is an example of this alternation: effort and rest, ascent and descent, always in the same river.

This is where Taoism approaches pantheism and panpsychism. If every reality exists only in relation to its opposite, every element participates in a cosmic thought or sensibility. Yin perceives yang, yang perceives yin, and this mutual "consciousness" permeates everything. For a Taoist master, even a motionless stone contains the rhythm of time and therefore an inner principle, invisible but active.

This unity of opposites has had a profound influence on contemporary visions of universal interconnectedness. In the language of modern physics, quantum entanglement describes particles that are linked to each other even at immense distances. The Tao would say that one particle 'knows' the other because they both belong to the same breath. In ecology, an ecosystem lives and thrives only if opposing forces (growth and decay, predation and reproduction) remain in balance.

The Tao, in its essence, is the space where every tension finds a rhythm. It is the art of seeing day and night as a single story. It is recognizing in the fullness of the moon the promise of its waning, and in the heart of a storm the calm that will follow. Through this lens, the world is a great organism that breathes through opposites.

Taoism and the living universe: panpsychist and pantheist suggestions

The Tao is not an abstract concept. The Tao is a living presence that vibrates in every element of the world, from the blade of grass that springs up in the silence of a valley to the star that burns in the interstellar void. In Taoist thought, even what appears inanimate—a rock, a drop of water, a grain of sand—participates in the same cosmic breath. Its existence is linked to a primary energy that pervades every level of reality.

This insight is surprisingly close to panpsychism. Every fragment of the cosmos, however small or immobile, possesses a form of sensitivity. This is not individual consciousness like that of humans, but a subtle feeling, a rhythm that connects it to the whole. As Zhuangzi (4th century BC) wrote, "*between heaven and earth, among the ten thousand beings, there is nothing that does not breathe together with the Tao*."

In classical Taoism, the value of an object is not measured by its usefulness. A rock is not important because it can be climbed or carved, but because it belongs to the universal balance. The masters regarded stones as guardians of a slow time, capable of teaching patience and stability.

Sacred mountains, such as Huangshan or Mount Tai Shan, were considered places where the Tao was concentrated with particular intensity. Their shapes, shrouded in mist, gave monks the perception of moving within a living landscape, where every rustle and every light changed like the breath of an organism. This was not a metaphorical phenomenon: for them, the mountain itself had a "heart," the wind had a "language," and water had a "memory."

Many travelers of the Tang dynasty recount ascents to Huangshan at dawn. There, in the silence broken only by the sound of footsteps on the stone paths, they felt that the sun, the rocks, and the clouds formed a single body. This experience fueled the belief that the world was a great organism, not a sum of separate parts.

From a pantheistic point of view, the Tao is the sum of all things, without any division between the principle that orders and that which is ordered. It is similar to Spinoza's vision of

Deus sive Natura: God and Nature are the same reality. For Taoism, there is no external 'architect'; law and matter are identical. The river does not follow a plan imposed by a distant will: the river is the law, and its current is part of the very structure of the universe.

This conception has a direct influence on modern visions of cosmic interconnectedness. In theoretical physics, hypotheses about quantum consciousness approximate this image of a universe in which every entity participates in a common field of information. In ecosystem biology, the concept of the living web echoes the Taoist idea that every element, even the smallest, contributes to the overall balance.

After all, when a Taoist monk meditates silently on a mountain, he is not escaping from the world. He is listening to the world. He is recognizing that his mind is part of a larger breath, shared with stones, with tree roots, with the light reflecting on the water. This is the life of the Tao: a cosmos that feels, reacts, and renews itself without ever ceasing to be a single body, infinite and indivisible.

Influence of the Tao on modern pantheistic and panpsychistic thought.

The Tao, in its original vision, is an immanent force. It is both principle and matter. This idea has not remained confined to classical texts such as the *Tao Te* Ching or the Zhuangzi. It has crossed centuries and borders, finding unexpected correspondences in European pantheistic thought and modern panpsychistic theories.

In the 17th century, Gottfried Wilhelm Leibniz became interested in Chinese philosophy thanks to Jesuit missions. Reading descriptions of the Tao, the philosopher saw an affinity with his conception of pre-established harmony. In his letters, he noted that the Chinese knew a 'law of heaven' that governed opposites in balance without external intervention. For Leibniz, as for the Tao, the universe was a network in which every monad, even the smallest, reflected the entire cosmic order.

Two centuries later, European Romanticism found in the Tao a poetic model of union between man and nature. Writers such as Novalis and Goethe admired Eastern texts for their ability to dissolve the separation between subject and landscape. In his youth, Goethe copied verses from *the Tao Te Ching* and integrated them into his reflections on the "living totality" of nature.

In the 20th century, biologist James Lovelock, in formulating the Gaia hypothesis, unknowingly took up a Taoist insight. In Lovelock's view, the Earth is not a container of life, but a living organism that self-regulates its own conditions. The analogy with the Tao is clear: the planet breathes, reacts, and adapts as a single body would. Lovelock collaborated extensively with microbiologist Lynn Margulis, who argued that microorganisms and sedimentary rocks also participate in this great biological balance.

On a scientific level, some interpretations of quantum physics evoke the image of the Tao. The concept of entanglement describes particles linked together even at enormous distances, as if they shared an "invisible bond." Modern physicists recognize that this connection cannot be explained by local interactions: it is an intrinsic property of the system. The ancient Tao would have spoken of "a thread that unites heaven and earth."

Baruch Spinoza's pantheistic philosophy, with its Deus sive Natura, finds a natural ally in the Tao. In both visions, principle and manifestation coincide. There is no God external to matter or separate from time. The flow of the river, the cycle of the seasons, the planetary motion are not the products of an external design, but living parts of the same eternal substance.

Historical anecdotes show this continuity of perception. In 1919, the French poet Paul Claudel, ambassador to China, wrote after a visit to Mount Emei Shan that the landscape appeared to him to be "self-aware." The clouds breaking over the peaks, the outline of the gorges, the sound of the water, all created a single presence. Claudel did not use the term "Tao," but his description resonated with Taoist sensibility, that of a world that perceives and expresses itself.

The Tao, from a contemporary perspective, is also an ethical invitation. It asks us to perceive the world as a living structure, devoid of absolute hierarchies. Man, stone, wind, and stars share the same origin and the same destiny of transformation. The Way, the silent matrix of all things, does not impose order from above. It generates order from within, through the constant relationship between forces.

For this reason, many modern thinkers of panpsychism see in the Tao an ancient confirmation: the universe is not an indifferent machine, but a plural subject, sensitive down to the last speck of cosmic dust. In this vision, science and metaphysics touch each other. Matter is alive, and life is matter that recognizes itself.

Brahman: the cosmic breath of Vedic India.

At the heart of the Vedic tradition, the concept of Brahman presents itself as an unlimited horizon. It is the ultimate reality, impersonal and infinite. It is not an external creator, but the very fabric of the universe. River, light, stone, air: every sentient and insentient form is a manifestation of Brahman. This vision has spanned eras and languages, influencing pantheistic philosophies and panpsychist theories up to the present day.

The ancient Upanishads speak of a total unity where the individual self, *the Atman*, coincides with Brahman. "*Tat tvam* asi" (*You are that*) is perhaps the most radical phrase ever uttered about the relationship between consciousness and the world. Nothing is excluded. Human beings are not the privileged center. Animals, mountains, and planets share the same essence.

Western thought has recognized a powerful model in these ideas. Arthur Schopenhauer, studying translations of Sanskrit texts in the 19th century, saw in Indian metaphysics the confirmation that will and reality were a single universal current. Baruch Spinoza, without direct contact with the Vedas, used a similar formulation in his "Deus sive Natura," imagining God and Nature as two names for a single substance.

In the 20th century, Hinduism influenced ecological and cosmological thinking. The belief that the Earth is the living body of Brahman found correspondence in James Lovelock's Gaia hypothesis and in the insights of Fritjof Capra. Here, matter is not passive: it breathes and interacts, just as in the ancient Vedic formulas.

Cultural exchanges amplified this influence. Swami Vivekananda, at the Parliament of Religions in Chicago in 1893, described Brahman as "*the shore-less ocean, where every being is a wave.*" Those words struck thinkers such as William James, paving the way for a dialogue between Indian metaphysics and the philosophy of consciousness. Later, Alan Watts and other Western interpreters reinterpreted Brahman in a modern key, linking it to quantum physics, the concept of cosmic energy, and theories of diffuse consciousness.

This brief introduction shows how Hinduism has offered pantheistic and panpsychistic thought a language and model that remain relevant today. Not as a spiritual relic, but as a living structure for interpreting the world: a universe where every fragment, even the most insignificant, is part of an infinite reality that has no beginning and no end.

Brahman as cosmic unity and pantheistic model.

In *the Vedas*, composed between the 2nd and 1st millennia BC, Brahman is revealed as the absolute principle. It is not a god located in space, nor is it a celestial ruler with capricious will. It is the indivisible unity that sustains and permeates everything. The chanting of a priest during sacrifice, the slow flow of a sacred river such as the Sarasvati, the motion of the constellations in the Indian night: everything belongs to a single, boundless order.

This unity is not just an abstract concept. In the Upanishads, the phrase "*Tat tvam* asi" (*You are that*) becomes a bridge between the human microcosm and the universal macrocosm. The Atman, the deepest self of man, is identical to Brahman. This means that reality is not divided. The fragment and the whole, the wave and the ocean, have the same substance.

European thinkers welcomed this insight with a mixture of amazement and recognition. In the first half of the 19th century,

Arthur Schopenhauer studied the translations of Sanskrit texts produced in those years by German and English scholars. He recognized in Brahman the confirmation of his idea of a world as th l manifestation of a single, blind, and universal will. The same conceptual paths, traveled centuries earlier by Baruch Spinoza, led to a surprisingly similar formulation. In Deus sive Natura, Spinoza described God and Nature as two names for a single infinite substance, where "*eternal order*" is not external creation, but immanent essence.

The correspondence between these worlds of thought is also evident in cultural details. In an Indian village, a Brahmin reciting verses from *the Rig Veda* at sunset is not worshipping a separate entity: he is reaffirming that sunlight and human breath have the same root. Similarly, Spinoza, in 17th-century Amsterdam, wrote that *'all that is, is in* God' and that nothing can exist outside this unity.

Brahman, as a pantheistic model, offers a radical view of the universe. It leaves no room for the divide between sacred and profane, between spirit and matter. Everything is an expression of the same ultimate reality. A neem tree in a courtyard in Benares, a noisy market in Calcutta, the quiet of the Nilgiri hills: every place, every form, is woven into the same cosmic fabric.

This conception has made it possible to think of the divine not as exclusive but as inclusive. It has paved the way for philosophies which, although distant in language and tradition, have recognized in it a universal principle: the world is not a collection of disjointed parts but a unified organism. And in the unity of Brahman lie, even today, the roots of much pantheistic philosophy and modern panpsychistic interpretations.

Diffuse consciousness and panpsychist suggestions

In the Vedic view, consciousness is not a privilege of the human mind. It is a universal quality. Brahman does not live only in humans or gods, but penetrates stones, rivers, wind, animals, clouds, and stars. In the songs of the Rig Veda, the mountain is an ancient being, the river is a mother who gives life, and the air is cosm l breath. Even what appears inanimate is endowed with a 'subtle life', invisible but real.

The ancient rishis saw the world as a conscious fabric. Every particle, from a grain of sand to sunlight, participates in a sentient principle, albeit at different levels. This vision dissolves the clear boundary between subject and object. There is no 'outside' devoid of consciousness: there is only deeper or more manifest consciousness.

In the 20th century, some schools of thought took up this insight under the name of panpsychism. Consciousness, according to this perspective, is an intrinsic property of matter. Physicist David Bohm described the universe as an "implicate order" in which each part contains the whole. Writing that the fragment and the whole are inseparable, Bohm formulated an idea that, if expressed in Sanskrit, would have found a place in the verses of the Upanishads.

A spiritual master from Punjab might have smiled upon hearing a quantum physicist speak of "*information distributed everywhere*." For classical Hinduism, the sound of a drum in a village and the luminous vibration of a star belong to the same field of consciousness. This field, according to Vedic philosophy, is Brahman.

Some historical episodes show the fascination exerted by this conception. When Swami Vivekananda, at the Parliament of Religions in Chicago in 1893, declared that '*every atom of the cosmos is imbued with infinite* consciousness', listeners were struck by a statement as radical as it was universal. For many Western thinkers, these words were a gateway to a broader understanding of the relationship between mind and matter.

Alfred North Whitehead's philosophy, with its concept of "occasions of experience" present throughout reality, is also influenced by this matrix. In his view, every entity possesses a margin of self-experience. In Vedic terms, every fragment is a point where Brahman is present.

Recognizing a diffuse consciousness changes one's perception of the universe. One no longer observes a scene from the outside: one is part of the scene. The starry sky is not an inert backdrop, but a chorus of presences. A banyan tree is not a plant structure, but an ancient guardian in which Brahman breathes.

Hinduism thus provides modern thought with a powerful image: a living universe, made up of countless eyes, observing from different forms. This suggestion, inherited from texts written thousands of years ago, remains one of the most daring ideas in the dialogue between philosophy, spirituality, and science.

Ecological ethos and universal interdependence

In Vedic thought, nature is not a commodity to be disposed of. It is part of the very body of Brahman. Destroying a forest, draining a lake, or polluting a river does not only cause material damage: it is a wound inflicted on cosmic unity. Ancient texts remind us that "what lives is Brahman" and that every creature, visible or invisible, is a unique knot in the universal fabric.

The Vedic priest who prepares a sacrifice does so not to appease a separate deity, but to restore harmony between man and the world. Even everyday acts, such as collecting water from a sacred river, are accompanied by ritual formulas that recognize the vital value of the element. In this view, violating the natural balance means breaking the law that sustains the cosmic order.

This perception has reemerged, in new forms, in contemporary ecological thought. The Gaia hypothesis, proposed by James Lovelock in the 1970s, describes the Earth as a living organism capable of self-regulation. It is an image that resonates strongly with Vedic philosophy: a planetary body that breathes, feeds, expels toxins, and maintains an internal balance. Fritjof Capra, in *The Web of Life* (1996), states that every biological system is inter d with others in a dynamic network. This network, as seen by an ancient Brahmin, would be a visible manifestation of Brahman.

The parallel is not forced. In Rajasthan, the Bishnoi communities have protected the fauna and flora of the desert for centuries, following religious precepts that prohibit the felling of trees and the killing of animals. In 1730, the young Amrita Devi opposed the destruction of a khejri forest by the Maharaja's soldiers. She embraced the trunk and declared that "*a tree is our life*." She was killed along with hundreds of other inhabitants, but their action made the forest sacred and prevented its destruction. This episode, read in a Vedic key, is a concrete example of the defense of the body of Brahman.

Universal interdependence, in Vedic thought, is not a theoretical concept. It is a vital law. Human beings are not masters of the Earth, but temporary custodians. Every action has an impact on the whole. Burning a field without respect, diverting the course of a river, eliminating an animal species: all these acts result in imbalances that reverberate throughout the cosmic cycle.

In a modern world facing climate crises and biodiversity loss, this ancient awareness offers a powerful ethical perspective. Living in harmony with the web of life is not romanticism, but a necessity for survival. Brahman, in this reading, becomes an ecological principle: remembering that 'what lives' is never separate from us.

Cultural encounters and the rebirth of unitary thought

At the end of the 19th century, the journey of some key figures marked a decisive moment in the dialogue between East and West. Swami Vivekananda, a young Bengali monk, crossed the Atlantic to attend the *Parliament of Religions* in Chicago in 1893. On the stage of the large auditorium, surrounded by theologians, philosophers, and free thinkers, he spoke of Brahman as "*the boundless ocean*" and of every living being as "*a temporary wave destined to return to the original sea.*" His tone was calm, but his image possessed a poetic force capable of leaving a lasting mark.

American transcendentalists, who already read Emerson and Thoreau, found in Vivekananda's words a bridge to a spirituality that united individual freedom and cosmic belonging. Some Harvard scholars, including William James, noted how these ideas could dialogue with pragmatism and the nascent psychology of consciousness. James wrote that "*contact with Eastern philosophies broadens the concept of experience*," recognizing that consciousness could not be reduced to the individual mind alone.

In the 20th century, the flow of encounters intensified. With the arrival of masters such as Paramahansa Yogananda, author of *Autobiography of a* Yogi (1946), the West discovered the possibility of learning not only metaphysical concepts, but also meditation and yoga practices capable of directly perceiving the interconnectedness of all things. The pantheistic idea ceased to be a distant theory: it became a bodily and sensory experience.

During those years, figures such as Aldous Huxley also engaged with Vedic thought. Huxley defined the "perennial philosophy" as a set of insights shared among the great spiritual traditions, in which Brahman occupies a central place. For him, unitary thought was not a dogma but a network of correspondences, from Sufi mysticism to Zen Buddhism, from Neoplatonism to contemporary physics.

In the 1960s and 1970s, philosopher Alan Watts brought a non-dogmatic reading of Eastern visions to Western debate. With his ability to communicate directly, Watts translated Vedic concepts into language understandable to an audience that studied Heisenberg and listened to the Beatles. In his speeches, Brahman became an "infinite network of relationships," an image that spoke as much to those meditating in an ashram as to those working in a physics laboratory.

Places such as *the Esalen Institute* in California became crossroads for this cultural encounter. Seminars were organized here that combined neuroscience, Vedic spirituality, and phenomenological philosophy. Scientists such as Francisco Varela participated in these conversations, exploring how meditative experience could contribute to the understanding of the conscious mind.

The dialogue was never a mere academic exercise. Many researchers, drawn to the Vedic vision, began to consider consciousness as a widespread property of the universe, recovering insights from ancient panpsychism and linking them to the discoveries of quantum physics. The idea that "*everything breathes*" was no longer a poetic metaphor: it became a hypothesis to be tested.

Today, in a global world, the words spoken by Vivekananda in Chicago in 1893 still resonate. The "boundless ocean" is not only a spiritual image; it is also a scientific metaphor for a universe in which every fragment is interdependent. Unitarian thought, reborn thanks to these cultural encounters, continues to move between the university lecture hall and the meditation hall, with the same breath that, for Vedic philosophy, belongs to Brahman.

Brahman, with its infinite and immanent nature, is still today a bridge between philosophy, science, and spirituality. It speaks of a living universe, in which consciousness and matter are not separate entities but aspects of the same principle. In this perspective, modern panpsychism has ancient roots, and pantheism receives a millennial confirmation: what exists, lives; and what lives is part of the same ultimate, silent, and omnipresent reality.

The harmony of the Mandate of Heaven and cosmic cyclicality.

Imagine a time when the power of a ruler was measured not only by the strength of his armies or the vastness of his territories, but by his ability to live in harmony with a higher principle: the Mandate of Heaven.

In ancient Chinese culture, this Mandate (in Chinese Tiānmìng) was seen as the seal of legitimacy granted by Heaven, the force that ordered all things. It was not a random gift: it was bestowed upon those who demonstrated virtue, righteousness, and the ability to maintain balance between the human and natural worlds. Governing according to the Mandate of Heaven meant ensuring that the laws of men reflected the laws of the universe.

The 'harmony of the Mandate of Heaven' represented the highest point of this vision: a time when the rhythm of the seasons was favorable, the earth yielded abundant fruits, and the people lived in peace, because those who held power did so with justice and benevolence. But this divine favor was not eternal: if the ruler fell into corruption, Heaven withdrew the Mandate and history opened up to new changes, often marked by revolts or dynastic transitions.

From a philosophical point of view, the Mandate of Heaven occupies a central place in Confucianism. For Confucius (Kǒng Fūzǐ, 5th century BC), it was a moral commitment: a constant reminder to rulers to be examples of virtue, to live in justice (yì) and benevolence (rén), so as to ensure social harmony.

Taoism, although less interested in politics, recognized the Mandate of Heaven as a reflection of the natural order: a sign that the Dao, the "Way," flowed unhindered and that the rhythms of the cosmos were in harmony with human actions. In this view, harmony was not only a balance between ruler and subjects, but also an inner harmony between the individual and the invisible fabric of the universe.

Historically, the concept originated and became established during the Zhou dynasty, long before Confucius, but it was he who imbued it with universal ethical significance. After him, other thinkers and schools gave it different interpretations, but the core remained unchanged: power is legitimate only if based on virtue and harmony with Heaven.

According to ancient texts, when a ruler abuses his position, the Mandate is withdrawn. The withdrawal does not take place with a proclamation, but manifests itself through natural disasters, popular uprisings, or economic crises. The fall of the Shang dynasty, recounted in *the "Book of Documentation"* (*Shu Jing*), is presented as a direct consequence of this rejection by Heaven.

The most interesting element, from a pantheistic point of view, is the vision of a moral universe. In the Mandate of Heaven, natural law and ethical law coincide. A flood or famine are not random events. They are signs of an imbalance between human action and the cosmic rhythm. In this sense, the concept is close to certain Buddhist interpretations, where karma is an expression of a causality that intertwines actions and conditions in the world. Buddhism, which arrived in China in the 1st century AD, found fertile ground in this idea: both systems see the cosmic order as something to be respected, not dominated.

Confucius reinterpreted the Mandate of Heaven in terms of everyday ethics. He asserted that a ruler must cultivate virtue, since it is through righteousness that he 'serves Heaven'. The justice of a government thus becomes a reflection of universal justice. It is a concept that over time influenced not only politics but also art and literature. The painter Ma Yuan, in his views of mountains shrouded in mist, sought to show the harmony between microcosm and macrocosm, between man and the breath (qi) that permeates all things.

Anecdotes from Chinese imperial history illustrate this connection well. In 1644, when the Ming dynasty collapsed, chroniclers of the time noted that earthquakes, famines, and abnormal weather phenomena had occurred before the fall. For the mentality of the time, these were proofs that Heaven had withdrawn its mandate. The people read a moral response in the signs of nature.

The common elements with Buddhism are also found in cyclicality. Buddhism sees history and life as a flow of births and deaths, dissolutions and rebirths. The Mandate of Heaven moves in a similar way: dynasties rise, flourish, decline, and are replaced in a cycle that mirrors the continuous transformation of the natural world. This link between politics, nature, and spirituality creates a unified way of thinking, in which the destiny of a society is intertwined with the breath of the universe.

Today, some scholars of comparative philosophy are reinterpreting Tiānmìng in the light of planetary ecology. If modernity, with its environmental crises, shows signs of global imbalance, it is possible to see in this condition a metaphor for a withdrawn Mandate of Heaven. Not as punishment, but as a natural consequence of a departure from the laws of interdependence. It is a reflection that speaks to the present with surprising relevance, inviting us to rethink the relationship between political action, ecological balance, and cosmic rhythm.

Buddhism.

In Buddhism, there are several elements that can be read as panpsychist or pantheist suggestions, and which have had a direct or indirect influence on the development of a unitary pantheist or panpsychist thought, especially in comparative contexts and in modern reinterpretations.

The principle of universal interdependence (Pratītyasamutpāda).

In a monastery in Sarnath, a few kilometers from Varanasi, an elderly monk explained a simple yet dizzying truth to a group of novices: "*Nothing exists for itself. Everything is connected to everything else.*" This sentence summarises the doctrine known in Sanskrit as Pratītyasamutpāda, often rendered as "conditioned co-production" or "dependent origination".

Buddhist thought places it at the center of its worldview. According to this doctrine, every phenomenon arises because of a network of conditions, and disappears when these conditions cease. A seed germinates because it finds soil, water, light, and heat; but none of these elements, taken alone, can generate a plant. It is their invisible and complex intertwining that gives rise to life.

This perspective overturns the idea of a universe made up of independent objects. In Pratītyasamutpāda, every fragment of reality is part of a single, continuous fabric. The Buddha expresses this insight in the dialogues reported in the Saṃyutta Nikāya: "*With the arising of this, that arises; with the cessation of this, that ceases.*" The lesson is clear: nothing arises by itself, nothing remains without connections.

For a pantheist, this view suggests that the universe is a single living body, where every part participates in the same substance. For a panpsychist, it opens up the possibility that this shared network is not only material but also experiential: if everything is related, then every entity participates in some way in a process of diffuse consciousness.

In the 13th century, the Japanese Zen master Dōgen reinterpreted the doctrine of interdependence with powerful images. In his work Shōbōgenzō, he wrote that every single grain of dust contains the entire sky. This is not a poetic metaphor, but a literal vision of universal interpenetration. Surprisingly, this idea is close to certain developments in contemporary physics: quantum mechanics describes a world where particles remain correlated ("entangled") even at immense distances.

A concrete example helps to understand this better. In rural villages in Bhutan, rice cultivation follows a lunar calendar. Farmers do not see the moon as a mere celestial body, but as part of the life cycle that unites water, plants, and human activity. A change in rainfall affects crops, which in turn determine social and religious life. Every element—celestial, terrestrial, human—is part of the same balance.

Pratītyasamutpāda is not only a philosophical observation but also an ethical guide. If nothing lives in isolation, damaging one part of the system is equivalent to damaging the whole. It is an idea that anticipates modern ecological sensibilities, where protecting a forest is not only about protecting trees but also about conserving a planetary organism.

Even from a comparative perspective, the principle of interdependence has dialogued with Western currents. In the 17th century, Baruch Spinoza formulated the famous equation "Deus sive Natura": God is Nature, and Nature is God. The whole is the One. The comparison with Buddhist thought is not forced: both describe a cosmic order without radical hierarchies, where separation is illusory.

Today, scholars such as the American philosopher Evan Thompson have used Pratītyasamutpāda as an interpretative key for theories of distributed consciousness. In an interconnected universe, consciousness is not the exclusive property of the human brain, but can be seen as a process emerging from broader relationships, perhaps extending (as some panpsychist hypotheses suggest) to all of reality.

Thus, from the Buddha's preaching in the forests of northern India to the contemporary debate between philosophy and neuroscience, the principle of interdependence remains a map for navigating a world where nothing is isolated. A world, we might say, where knowing means recognizing the invisible bonds that unite us to the whole.

The nature of Buddha (Buddha-dhātu)

One winter morning, the Chinese master Huineng, sixth patriarch of Chan, heard a monk reciting a sutra on the power of the enlightened mind. Huineng interrupted him with words that have remained famous: "The original mind is already pure. It should not be sought elsewhere." These few sentences encapsulate the heart of the Buddha-dhātu doctrine: every being possesses Buddha nature, that is, the intrinsic potential to awaken to the truth.

In Mahayana schools, this principle is not a poetic concession, but an ontological statement. *Tathāgatagarbha* (*the womb of the Buddha*) is the spark that dwells in all living beings. It is not a physical object, but a universal principle that runs through all forms of life, like the sap that invisibly flows through the roots, branches, and leaves of a tree.

This idea, when read from a panpsychist perspective, opens up intriguing scenarios. Consciousness, understood as the ability to perceive and react, would not be the prerogative of the human mind. Rather, it would be a widespread principle, a fundamental and l quality of being, capable of manifesting itself in complex organisms but also—according to symbolic and philosophical interpretations—in the inanimate world, as a reflection of the Dharma itself.

The Śrīmālādevī Siṃhanāda Sūtra, one of the key texts of Mahāyāna, states that the nature of Buddha is "permanent, stable, eternal, and immutable." This phrase has given rise to divergent interpretations over time. The scholars of Nālandā saw it as an ethical message: even the most tormented being has the possibility of awakening. Contemporary philosophers, such as the Japanese Masao Abe, have proposed a less anthropocentric view, where this 'nature' is part of the entire structure of the universe.

An anecdote from Zen master Shitou Xiqian illustrates the meaning of this doctrine well. During a walk in the mountains, Shitou stopped in front of a large boulder and said to his students: "*This stone is already Buddha, but it does not know it.*" The phrase, which is only paradoxical in appearance, suggests that the sacred is not confined to biological life but permeates all matter, in a continuum that nullifies the separation between conscious and inert.

In the modern era, the ecological rediscovery of these insights has led us to consider Buddha nature as a powerful metaphor for our relationship with the planet. If every being—from humans to the smallest insect—is the bearer of the same fundamental nature, then harming one part of the biosphere means offending the whole. This idea finds parallels in Spinoza's pantheistic vision, where every fragment of existence is an expression of a single infinite substance.

Comparatively speaking, the concept of Buddha nature is reminiscent of certain formulations in ancient Greek philosophy. The Stoics, for example, spoke of pneuma as a vital principle pervading everything, from the human body to the stars. In both cases, it is an immanent energy that confers order and life, and which admits no exclusions.

Places steeped in meaning help to give substance to this vision. In Bodh Gaya, where according to tradition the Buddha attained enlightenment under the sacred fig tree, monks, , and pilgrims meditate in the belief that the same potential can sprout in everyone. The temple therefore does not celebrate a unique event, but a universal archetype.

From a panpsychist point of view, the Buddha-dhātu suggests a radical paradigm: consciousness is not the result of recent evolution, but an original principle present from the beginning, which matter and life manifest in different forms. Some theoretical physicists, such as David Bohm, while operating on a secular level, describe reality as an 'implicate order', where each part contains the information of the whole. The resonance with the idea that 'every being is already Buddha' is evident.

Thus, from the mountains of China to the temples of India, to contemporary philosophical speculations, the doctrine of Buddha nature continues to challenge the frontiers of thought. It delivers a universal message: beneath the differences in form and function, there is a shared quality that unites us with the rest of creation. A message that, in panpsychist readings, finds a natural bridge with the idea that the cosmos itself is, in some way, conscious.

The concept of Śūnyatā and the "full emptiness" .

In a cave on the Ladakh plateau, a Tibetan monk recites verses from the Prajñāpāramitā Sūtra in a low voice. Snow falls slowly outside the door, the sound rhythmic like breathing. One phrase recurs several times: *'Form is emptiness, emptiness is form'*. This is the essence of Śūnyatā, the doctrine of emptiness.

The term does not indicate a meaningless "nothingness." Rather, it expresses the absence of independent existence. Every phenomenon is devoid of an immutable self. The mountain exists thanks to the erosion that shapes it, the wind that brushes against it, the light that reveals its form. Without these relationships, the "mountain" as a separate entity has no meaning.

The philosopher Nāgārjuna, in the 2nd century CE, articulated this insight in the Mūlamadhyamakakārikā. He wrote that all concepts and things are "*empty of* themselves" because they depend on causes and conditions. Emptiness does not destroy reality, but reconfigures it as a dynamic network.

In some modern interpretations, Śūnyatā becomes like an infinite "field" of possibilities. A field where anything can manifest because nothing is fixed. It is compared to the ocean of Brahman in Hinduism, the single principle from which everything emerges. In science, David Bohm's description of the "implicate order" offers a fascinating parallel: a deep and indivisible level of the universe, where each part contains the information of the whole.

This vision favors pantheistic reinterpretations. If every phenomenon exists by virtue of the relationships that run through it, then the universe itself is an indivisible organism—a whole that breathes in infinite forms. For a panpsychist, this opens up a further horizon: the "void" is not absence, but the power of diffuse consciousness, a reservoir from which the experiences of all beings arise.

A Zen anecdote illustrates the existential dimension of this insight. When asked what emptiness was, Master Kōdō Sawaki pointed to the sky and said, '*There is nothing to grasp, yet everything is* here'. The sky is not a passive container, but the space that allows everything to exist.

In the Tibetan tradition, Dzogchen meditation practices seek to perceive this emptiness as "luminous fullness." Not a metaphysical black hole, but a state in which every form is seen as transitory and open, never isolated. It is precisely this openness that makes reality creative and interconnected.

From a comparative point of view, Śūnyatā is reminiscent of the concept of kenosis in Christian mysticism: the "emptying" of oneself to make room for the divine. Here too, emptiness is not deprivation, but a condition for a greater fullness. Similarly, in quantum physics, the "vacuum" of the quantum field is overflowing with fluctuations and potentialities, and can generate particles from apparent nothingness.

Sacred places such as Mount Kōya in Japan embody this philosophy in everyday life. Here, monks move among ancient trees and silent temples, convinced that every step is supported by an invisible network of dependencies and possibilities. The 'full emptiness' manifests itself in the awareness that one's existence is woven together with that of others, human and non-human.

In short, Śūnyatā is a concept that disarms certainties and invites us to perceive the universe not as a collection of fixed things, but as a living fabric. It is an invitation to recognize that "emptiness" is the engine of form, and that this openness reflects the ultimate unity of reality. A perspective that, in panpsychist reinterpretations, sees this emptiness as the cosmic womb from which consciousness itself springs.

's historical influence on pantheistic and panpsychist thought remains.

In the 19th century, Arthur Schopenhauer found in Buddhism a surprising confirmation of his own insights. He studied the Dhammapada and the Mahayana texts translated into German by scholars such as Burnouf. In them, he found a worldview that dissolved the idea of an independent self. For him, this perspective was consistent with the notion of a "universal will" that permeates all phenomena, from the most complex organism to seemingly inert matter. Schopenhauer wrote that the truth of the Buddha coincided with the understanding that "*the world is a dream of the will.*" Here, the will takes on a panpsychist tone: it is not a human attribute, but the living, unconscious substance of existence.

In the 20th century, Buddhist thought entered into direct dialogue with Western philosophy and science. Daisetz Teitaro Suzuki, a Japanese Zen master, presented Buddhism in lectures and held in the United States and Europe. In the 1950s, he spoke at Columbia University about a "centerless" consciousness, comparing it to the dynamic structure of energy fields in physics. Suzuki related the Zen experience to Husserl's phenomenology, where the world is defined by relationships of perception rather than isolated substances. In this dialogue, Buddhist emptiness and the idea of diffuse consciousness took on a pantheistic value: the universe as an indivisible fabric in which each part reflects the whole.

In the United States, the transcendentalism of Ralph Waldo Emerson and Henry David Thoreau anticipated this interest. Emerson read Indian and Buddhist texts and found in them the principle of radical interconnectedness between nature and spirit. Thoreau, who studied the *Lotus* Sutra, wrote in his notes that "*the air I breathe is also the breath of the forest.*" These authors influenced generations of intellectuals, paving the way for links between ecology and spirituality.

In the 1960s, the Californian counterculture openly adopted Buddhist concepts. Communities such as Esalen in Big Sur hosted seminars on meditation and cosmology. Figures such as Alan Watts translated Śūnyatā into language understandable to a Western audience attracted to theoretical physics and evolutionary biology. Watts described emptiness as "the field on which all forms dance," a concept that accorded with the panpsychist views emerging in the philosophy of mind at that time.

This cultural trend had a decisive influence on the environmental movement. Meditation and the perception of universal interconnectedness moved from meditation halls to environmental campaigns. Activists such as Gary Snyder, a poet and Zen practitioner, spoke of the forest as a "conscious body," directly echoing Buddhist ideas of interdependence. Living in the Sierra Nevada mountains, Snyder saw trees, rivers, and animals as participants in the same life, an interpretation that is close to Spinoza's pantheism and contemporary panpsychism.

Today, in academic discussions of the mind-matter relationship, Buddhist doctrines of interdependence, fundamental nature, and emptiness continue to provide alternative models. Their historical influence is not limited to spirituality, but touches on the philosophy of consciousness, cognitive science, and quantum information theory. Many thinkers see Buddhism as a bridge between ancient insights and new scientific formulations: the cosmos as a large organism in which every particle of matter is, to some extent, aware of the network that sustains it.

Ecological and cosmic parallels

In a monastery nestled in the forests of Bhutan, monks begin their day with an ancient ritual: offering clear water in bronze cups not only to humans, but also to the mountain spirits and invisible animals believed to inhabit the valley. This simple, silent gesture encapsulates a key principle of Buddhism: human beings are part of a living network that encompasses all levels of existence.

This network is not metaphorical. It is real in the logic of interdependence (Pratītyasamutpāda). Suffering arises from imbalance, from the illusion of separation. When this principle is applied to the natural environment, the ground becomes fertile for pantheistic and panpsychist interpretations: the world is not a set of resources to be used, but an indivisible organism with intrinsic value and, in the most radical visions, a form of widespread 'sensitivity'.

Gary Snyder, poet and Zen practitioner, wrote in the 1970s that *'the forest thinks as much as the village'*. This deliberately provocative phrase summarized an idea present in the sutras: every component of nature possesses a dignity that, , stems from its being part of the whole. An ancient Japanese tale tells of a monk who, walking in the woods, bowed before every tree. When asked why, he replied, "Here, everyone is a teacher."

The panpsychist perspective finds a natural correspondence in this. If consciousness is a widespread property of the universe, then it is not illogical to consider mountains, rivers, and forests as participants in some form of awareness. In recent decades, this approach has become intertwined with deep ecology, a movement that calls for a symmetrical relationship between humans and the environment.

On a cosmic level, Buddhist metaphysics embraces a universe without a clear boundary between the living and the inanimate. Mahāyāna texts describe "myriads of worlds" that coexist and influence each other. According to certain schools, every vibration of a thought propagates without limits, touching stars and planets. In some contemporary reinterpretations, this vision coincides with cosmological hypotheses of an interrelated universe, where distant systems remain linked by quantum connections.

Western culture has embraced this suggestion at specific moments in history. In the 1960s, the Californian community of Esalen hosted scientists and Zen masters who discussed the universal 'mindfield', a concept that built bridges between Śūnyatā, particle physics and cosmology. The idea that the void is 'full' of relationships was supported by scientific data and reinforced a pantheistic view of the cosmos.

Today, the planetary ecological challenge makes these insights more relevant than ever. Seeing humans as part of a living whole leads to an ethic that rejects the destruction of natural balances. Buddhist thought invites us to dissolve the barrier between 'us' and 'nature', to the point of recognizing that, as Master Thich Nhat Hanh says, *'to breathe is to breathe with all the forests of the world'*.

Myth as a sacred map of the universe.

In the great Asian traditions, myth is not a fantastic tale intended for entertainment. It is a sacred map. It tells how cosmic forces are arranged and how human beings must move within them. In Buddhist sutras, as in Vedic Indian texts, the events narrated indicate an invisible order in which everything has a precise position.

Sutra Buddhism, for example, presents the myth of the Wheel of Dharma. It is not a static symbol: it is a diagram of reality. Each spoke of the wheel represents a universal law of interdependence (Pratītyasamutpāda). Observing it is like reading an atlas of the mind and the cosmos. The wheel suggests that consciousness is not isolated but distributed, like a river flowing through all forms of life.

Many Buddhist tales, such as the story of Indra and his net, offer pantheistic images in which each knot reflects all the others. The net extends infinitely, and each reflection contains the entire universe. In the panpsychist reading, this metaphor indicates that every element of reality possesses a form of participation in the whole, a glimmer of collective perception.

Monks of the Tibetan tradition describe the mandala as a topographical map of the cosmos. It is not only a ritual art, but a geometric language for describing the connection between man, earth, and sky. Those who walk the mandala in rituals symbolically enter into a relationship with every dimension of existence, until the barrier between the self and the rest of the world dissolves.

The ecological value of these sacred maps becomes evident when read in a contemporary key. Lama Dodrupchen Rinpoche, in a 1987 speech, stated: "To *destroy the mountain is to injure a member of the universal body*." In this view, every natural element has a role in a cosmic organism in which matter and spirit coincide. It is a perspective fully compatible with Spinoza's pantheistic philosophy and modern panpsychist formulations.

In current readings, Buddhist cosmology enters into dialogue with theories of distributed consciousness. Scientists and philosophers see in ancient images—Indra's net, mandalas, the wheel of law—visual models of an interconnected universe. Quantum physics, with the concept of entanglement, provides a new language to explain relationships that, in myths, were narrated through symbols and poetry.

Ultimately, myth becomes a bridge between culture and nature. It is never confined to sacred literature: it influences how we cultivate, how we build, how we live. Entire villages in Bhutan or Sri Lanka are oriented according to geomantic patterns drawn from religious texts in order to maintain *"harmony with the breath of the world."* In these communities, the idea that the universe is a living organism is not a theory: it is a principle of daily life.

The cosmic tree of Mesopotamia: the bridge between heaven, earth, and the underworld.

In Mesopotamia, the image of the tree is never that of a simple natural element. It is a living axis, a vertical compass connecting the three regions of existence: heaven, earth, and the underworld. In the cylinder seals of the Assyrian and Babylonian periods, the "tree of life" often appears in the center, guarded by hybrid beings or rulers. The trunk rises towards a sky inhabited by stellar deities; the roots sink into the invisible realm, where primordial waters flow.

In the 9th century BC, the Assyrian king Ashurnasirpal II had a richly stylized tree with geometric branches and fruits carved into his residences in Nimrud. For him, that image was a political and spiritual program: the sovereign presented himself as the guardian of the cosmic order, capable of maintaining the balance between celestial and terrestrial forces. In a tablet found in Nineveh, the priest Ishme-Dagan writes that "*the tree breathes for all, and its sap is the covenant between the gods and men.*"

The symbolism of the tree in Mesopotamia is linked to the underground waters of Apsû, the cosmic reservoir from which life springs. The roots of the tree draw directly from that sacred source, while the branches intercept the light of Shamash, god of the sun and justice. In the pantheistic interpretation, the tree becomes an organism in which matter and spirit, day and night, the visible and invisible worlds merge. It is an image that can be interpreted as an ancient form of panpsychism: every part of the cosmos is animated, participating in a diffuse and universal consciousness.

Many Sumerian myths describe the tree as the home of female deities associated with fertility. In a hymn dedicated to Inanna, goddess of love and war, there is a story of a sacred tree planted on the banks of the Euphrates and cared for by the goddess with daily attention, "*so that the branches touch the dwelling place of the stars and the roots speak to the waters.*" This image combines a maternal gesture with cosmic balance, uniting mystical vision with agricultural practicality.

Archaeologists have noted how, in the temple cities of Uruk and Lagash, the layout of sacred spaces imitates the structure of the cosmic tree: the main temple at the center, the canals as roots, the towers as branches. It is a symbolic urban design that translates a map of the cosmos into stone. Architecture does not only design shapes: it teaches order, codifies a relationship between humans and the world.

Today, the symbol of the cosmic tree takes on new meanings. Scholars of cultural ecology emphasize how the Mesopotamian idea of an interconnected life axis anticipates modern h l concepts such as the biosphere. Just as in the ancient conception, each level of the universe influenced the others, today we know that the climate, forests, ocean currents, and human societies form an integrated system. What the Mesopotamians recounted with stone and myth is the same intuition: the world is a single body, and every part of it is alive.

Below is a comparative table that compares the symbolism of the cosmic tree in Mesopotamia, India, and China, highlighting the convergences of pantheistic and panpsychist meanings. This table highlights how the cosmic tree, while taking on different forms and details, remains in all traditions a sacred axis that transmits life, order, and consciousness. The idea that the roots and branches are connected to different levels of the universe is common and underlines the pantheistic view of the world as a unified organism.

Table2- The cosmic tree in great civilizations.

Civilization	Symbol/Myth	Structure and Function	Pantheistic / panpsychist meaning	Cultural and contemporary connections
Mesopotamia	Cosmic tree in Assyrian seals and reliefs	Central trunk connecting heaven, earth, and the underworld; roots in the Apsû, branches toward Shamash	Universe as a living organism with a continuous flow of energy; each level participates in a collective consciousness	Influence on the temple architecture of Uruk and Lagash; anticipation of concepts of the biosphere and interconnected systems
Vedic India	Ashvattha tree in the	Roots facing the sky, branches towards the	Reality as a reflection of the Absolute; each leaf	Symbol used in meditation and yoga; represents

Civilization	**Symbol/Myth**	**Structure and Function**	**Pantheistic / panpsychist meaning**	**Cultural and contemporary connections**
	Bhagavadgītā	earth; inverted compared to common trees	represents a conscious being	the interdependence between individuals and the universe
India (Hindu mythology)	Tree of life in Puranic stories	Divine tree in Indra's paradise, source of immortal fruits	The universe as the garden of the divine; matter and spirit inseparable	Archetypes reflected in art and sculpture; sacred protection of plants
China (Taoism and ancient mythology)	Fusang tree	Tree located in the east, home of the sun, in Chinese mythology	Consciousness diffused throughout nature; sun and life as parts of a single life cycle	Influence on traditional painting and imperial gardens; symbolic preservation of natural spaces
China (pantheistic myth)	Jianmu Tree	Tree connecting earth and sky, ladder of the immortals	Universe as a living vertical structure; each node is part of a cosmic order	Concept used in feng shui and architecture; respect for the points of "connection" between earth and sky

Mount Meru: the living axis of the cosmos.

In ancient India, Mount Meru is not just any mountain. It is the axis around which the universe revolves, the vital column that supports heaven, earth, and the underworld. In the Purāṇas and Vedic cosmologies, Meru appears at the center of an immense circular structure: continents, seas, and mountain ranges are arranged around it, like the petals of a giant sacred plant.

The texts describe Meru with symbolic precision. Its summit sparkles with gold because it is made of suvarṇa, the metal of the gods. Its four faces each have a different color: white, yellow, red, and black, signs of the cardinal directions and cosmic energies. Celestial rivers flow down its slopes, including the Ganges, which in mythical vision springs from the summit of Meru to bring life to mankind.

Mount Meru is the home of Indra, king of the gods, and other Vedic deities. Each of its terraces is home to a kingdom: higher up live the luminous beings, in the center is the earth inhabited by men, and lower down are the underground regions where spirits and Nāga serpents dwell. In this vision, the mountain is a living organism. The summit sees, the slopes listen, the roots converse with the deep waters. It is an image that embodies a true primitive panpsychism: all parts of the universe participate in a common consciousness.

An ancient commentary on the Śrīmad Bhāgavatam recounts that the sage Narada, traveling between worlds, stopped on the summit of Meru. From there, he simultaneously saw the sunrise on one continent and the sunset on another. The myth conveys a powerful idea: the center of the world is also the center of universal perception, a point where time and space bend.

Indian iconography has translated this concept into architecture. Tower temples, such as the Ananda Temple in Bodh Gaya and the large complexes of Khajuraho, imitate the shape of Meru. Each architectural level corresponds to a cosmic band. Climbing to the top is a ritual act that symbolically repeats the journey from the human to the divine plane.

Even real geography has attempted to give coordinates to the myth. Some ancient texts identify Meru with the central Himalayas, others place it in an invisible dimension, accessible only to enlightened beings. For the pilgrim, the search for Meru is not a physical journey, but an inner ascent.

From a pantheistic perspective, Meru is the very body of the cosmos. The mountain trunk is the spine of the universe, its slopes are the limbs, and its roots sink into the underground seas of . The gods do not live outside the mountain: they are the mountain, just as man, earth, and sky are different manifestations of a single reality.

Today, the symbol of Meru is also found in contemporary meditation practices and spiritual ecology studies. The idea of a 'vital axis' connecting all planes of existence is surprisingly close to the systemic models of modern science. Thus, the myth, narrated for millennia around sacred fires, continues to offer a working image of the world: a cosmic organism where every part, from a grain of sand to a distant star, has its own consciousness and place in the universal order.

Great civilizations have imagined the center of the universe in different forms, but always with a common structure: an axis that unites what is above with what is below, making each level part of a single living organism. Whether it is a sacred mountain, a cosmic tree, or a luminous column, the symbol of *the "axis mundi*" expresses a powerful pantheistic vision: the universe is not a collection of separate parts, but a single body in which each element has its own consciousness.

Mount Meru, with its orderly levels and its function as a link between the divine and the human, belongs to this great symbolic family. Around it, in other cultures, equivalent images arise: the Nordic Yggdrasil, the Chinese Jianmu, the Mayan Ceiba, the Mesopotamian cosmic tree. Each embodies the same universal idea, while expressing it in different languages and materials.

The table below relates these representations, showing how peoples distant in space and time have developed a sacred map of the universe based on verticality, connection, and the life that permeates everything. This table offers a global pantheistic overview, in which Mount Meru becomes the central reference point, but symbolically dialogues with the Mesopotamian cosmic tree, Yggdrasil, the Chinese Jianmu, and the Mayan Ceiba.

Table3- Cosmic axes in the great pantheistic traditions.

Civilization / Tradition	Symbol / Myth	Structure and Function	Pantheistic / panpsychist meaning	Contemporary cultural connections
Vedic and Puranic India	Mount Meru	Mountain at the center of the cosmos, with ascending levels: divine realms, human world, underground realms	The universe as a living organism; the mountain is the body of the gods and the earth together; all parts are conscious	Present in temple architecture; used as a symbol in meditation and sacred geography
Tibet and Himalayan Buddhism	Kailash as Meru	Sacred mountain considered the residence of deities and the center of cosmic mandalas	Axis mundi connecting planes of existence; every natural element is a bearer of consciousness	Ritual pilgrimages; concept integrated into yogic and tantric practices
Mesopotamia	Cosmic tree	Trunk connecting heaven, earth, and the underworld; roots in the Apsû, branches toward solar deities	The cosmos as an interconnected organism; energy and consciousness flow in every direction	Influence on temple urban planning and systemic visions of nature
Northern Europe (Norse mythology)	Yggdrasil	Ash tree connecting nine worlds; roots in the realm of the dead, trunk in the human	Every creature, visible and invisible, is part of a network of life and consciousness	Source of inspiration in ecology and fantasy literature

Civilization / Tradition	Symbol / Myth	Structure and Function	Pantheistic / panpsychist meaning	Contemporary cultural connections
		world, branches in the divine realm		
China	Jianmu tree	Vegetable column connecting earth and sky; ladder of the immortals	The universe as a living vertical structure, in which each "node" is a point of conscious energy	Integrated into feng shui and architectural symbolism
Mesoamerica (Mayan)	Cosmic Ceiba	Tree that supports the sky and connects the three levels of existence	Sky, earth, and underworld as parts of a single living being	Continuation of the symbol in agricultural rituals and traditional festivals

Conclusions.

Looking back at the voices and images of these civilizations, we can see a golden thread running through eras and cultures: a universal intuition of an organic cosmos, in which vital energy is not confined to a distant god, but is spread everywhere. Egypt showed us the Nile as a divine artery; Mesopotamia revealed a heaven and earth united in a single breath; China and India sang of the Tao and Brahman as eternal principles underlying everything. Even myth, in its poetry, traced maps of the universe understood not as neutral territory, but as a sacred body.

Today, the call of these ancient perceptions echoes in the philosophical debate on panpsychism and in science that investigates consciousness as a fundamental property of matter. It is as if contemporary thinkers were reaching out to that past, not out of romanticism, but to recover an integral vision: the idea that understanding the universe also means feeling it alive. After all, what the ancients suspected, and what we are perhaps beginning to rediscover, is that 'Deus sive Natura' is not merely a philosophical concept, but an immediate experience of belonging: we are part of a whole that thinks, breathes, and flows through us, as it always has.

IV. The Presocratics and the animated world.

An idea is born An idea that originated in the port of Miletus.

In 6th-century BC Greece, the coast of Miletus was a crossroads of trade and ideas. Phoenician ships and Egyptian merchants brought not only goods, but also stories about the sky, the earth, and the gods. In this climate of cultural exchange, the first philosophical reflections on the relationship between matter and life were born. The thinkers we now call the Presocratics sought a universal principle capable of explaining the cosmos without constantly resorting to myth.

Thales, the most famous among them, lived in this port town around 624-546 BC. Some anecdotes describe him as a practical and curious man: Aristotle recalls that Thales predicted an exceptional olive harvest and, to prove it, bought all the oil mills in advance, earning a fortune. But what is interesting here is his idea that 'everything is full of gods'. Thales does not speak of an anthropomorphic pantheon, but of a widespread vitality. For him, water is the origin and substance of everything, not only because it nourishes and envelops all forms of life, but because it is shaped by an internal force, a cosmic psyche.

Anaximander, born a few decades later (c. 610-546 BC), broadened the theoretical horizon. He traveled, studied cartography, and constructed one of the first maps of the known world. His original principle was not a concrete element, but *the ápeiron*, the infinite. This undefined and unlimited substance generates everything and reabsorbs it in a continuous process of birth and death. For Anaximander, life is not a privilege of humans or animals: everything has an internal order, regulated by a cosmic justice that rebalances the cycle of existence. Living beings emerge from the primordial sea and transform themselves according to the laws of this eternal principle.

Thales and Anaximander seem to share the first philosophical formulation of the idea of a diffuse psyche: consciousness or life force is not confined to a human body or a particular god, but permeates the whole of nature. The river flows not only because of gravity, but because it participates in a universal life; the wind blows as a manifestation of an invisible but active principle.

These insights, developed in a context where myth still coexisted with rational observation, would influence generations of thinkers. The image of an animated world, formulated on the shores of the Aegean, is the first step in a long journey that leads, centuries later, to Spinoza's vision: Deus sive Natura. An idea that, from the port of Miletus, continues to speak today of a living and conscious universe.

'Water' as a fundamental principle.

Thales saw water not only as a physical substance, but as a living and pervasive principle capable of explaining the origin and cohesion of the world. When he stated that "everything is full of gods," he was not referring to an anthropomorphic pantheon, but to a widespread life force, a psyche inscribed in the very structure of reality.

To understand his thinking, we need to imagine the context of Miletus in the 6th century BC: a city of ports and markets, where practical observations of the world intertwined with the mythical traditions of the Mediterranean and the East. Water was there, visible in every form of life: in the fertility of the fields, in the sea routes, in the cycle of rain that provided sustenance. Thales, who traveled and studied astronomy and geometry, saw water as a universal element that not only sustains life, but generates it.

The choice of water as a principle (*arché*) also had symbolic significance. In Egyptian myths, which Thales probably encountered during his travels, Nun (the primordial ocean-)

precedes creation. In Mesopotamian cosmologies, the waters of Apsu and Tiamat generate gods and worlds. Thales takes up this archaic intuition but takes it in a more rational direction: water is not only the home of the gods, it is the original substance from which everything derives and to which everything returns.

The concept of the animation of matter is crucial to his vision. If water is the source of everything, it must also contain the force that motivates movement, growth, and transformation. This means that objects and beings are not inert: they all share the same vitality, a manifestation of the universal psyche. Water thus embodies the idea of a living continuum that unites plants, animals, humans, and even the stars, since even the sky—with its rain and fog—participates in this principle.

In this sense, Thales anticipates the theoretical thread that will run through panpsychism over the centuries: matter is not passive, but is the bearer of consciousness or at least of a vital principle. His 'water' is not only physical, it is also metaphysical: a bridge between the sensible and the intelligible world, which continues to influence ideas about nature as an intrinsically animated reality.

Anaximander and his concept of 'apeiron'.

The transition from Thales' thinking to that of Anaximander marks an important shift: from a concrete and observable principle—water—to an abstract and unlimited principle—apeiron.

Thales identifies water as the arche, a physical element that he sees everywhere in life and nature. His conception is still rooted in sensory experience and tangible phenomena, but it already contains an idea of universal animation: 'everything is full of gods' means that everything has a vital breath.

Anaximander, on the other hand, makes a conceptual leap: the apeiron is not a specific material element, but something indefinite, unlimited, and eternal, from which all things arise and into which they dissolve. It is a broader cosmic principle, which cannot be identified with any perceptible quality, and which can therefore explain the variety and mutability of the world better than a single element.

Thales offers a vision in which the vitality of the cosmos is linked to a common substance. The animation of reality springs from the universality of water as the source of life.

Anaximander, on the other hand, suggests that the origin and vitality of the cosmos come from a formless principle, capable of generating opposites (hot/cold, dry/wet) and governing cycles of birth and destruction. In this framework, what animates the world does not depend on a specific matter, but on a force, a perennial "field" that underlies every manifestation.

CONNECTION WITH PANPSYCHISM.

In Thales' thinking, panpsychism is rooted in direct observation: the original element not only gives form and substance to everything, but is intrinsically alive. Consciousness or vitality are qualities of matter itself.

In Anaximander, universal consciousness—if interpreted in a panpsychist key—is even more widespread and 'disembodied', because the apeiron is not tied to a form or an element, but is the principle of possibility of every form. This broadens the idea of animation: not only is every fragment of matter alive, but the entire cosmic process is permeated by an eternal and active origin that never runs out.

From a philosophical point of view, this is an evolution from 'living matter' (Thales) to 'living totality' (Anaximander). The transition to the apeiron can be seen as an early intuition of a 'proto-metaphysical' thought where the vital principle is not bound to physicality, but is a universal constant — an idea that anticipates more sophisticated panpsychist visions, up to contemporary ones that speak of consciousness as a fundamental property of the universe.

Other pre-Socratic figures who contributed to the idea of a 'diffuse psyche' in the cosmos.

In addition to Thales and Anaximander, several other pre-Socratics developed ideas that approximate the concept of a diffuse psyche, i.e., the notion that the universe is permeated by a vital force or consciousness.

Anaximenes of Miletus (c. 586-528 BC) identified air (aér) as arché, not only as a physical element but as a vital principle ("*just as our soul, which is air, sustains us, so the entire cosmos is sustained by* air"). Air is seen as universal breath; everything is animated because it is imbued with this invisible force, in constant motion and transformation.

For Heraclitus of Ephesus (c. 540-480 BC), the fundamental principle is fire, understood as the original element, symbol of constant transformation; fire is guided by the Logos, a rational order within the universe. The Logos is often interpreted as a cosmic "mind"; if everything changes according to this principle, every element is part of a guided process, therefore imbued with meaning and life.

According to Empedocles of Agrigento (c. 494-434 BC), the fundamental principle of everything lies in creation and transformation through four roots (earth, water, air, fire) united and separated by the forces of Love (philia) and Hate (neikos). His theory can be linked to panpsychism in that it predicts that both matter and the forces that move it are eternal and vital. Everything participates in a dynamic relationship, as if animated by affective and voluntary properties.

Anaxagoras of Clazomenae (c. 500-428 BC) introduced Nous (Intellect) as a fundamental principle. Nous is an ordering force that permeates and moves matter without ever completely mixing with it. Nous is linked to panpsychism in that it can be seen as a form of universal consciousness that governs cosmic processes, anticipating ideas of a 'diffuse mind' in the universe.

Finally, the Pythagoreans belonging to the school of Pythagoras (6th-5th century BC) postulate the cosmos as a fundamental principle. The Cosmos is numerical harmony; everything vibrates according to mathematical proportions. Here too, the possible connection to panpsychism is evident: if harmony is the hidden structure of the universe, it means that every part of the cosmos participates in a common "music," a sort of mathematical soul of the world.

All these figures, while differing in detail, share the belief that reality is pervaded by an active principle—be it air, fire, intellect, or harmony—capable of animating all that exists. Their legacy paved the way for subsequent panpsychist interpretations, in which consciousness is not seen as an exclusive property of humans but as a universal characteristic.

The following table represents a chronological reconstruction of the evolution of ideas, from the Ionian school to Plato and Aristotle, with particular attention to connections with historical panpsychism.

Table 4—Chronological reconstruction of ideas, from the Ionian school, connected to panpsychism.

Period	**Philosopher / School**	**Principle (arché)**	**Vision of the diffuse psyche**	**Connection with panpsychism**
6th century BC	Thales	Water	Life is intrinsic to everything that exists, animated by a universal life force.	First intuition that matter possesses a living principle.

Period	**Philosopher / School**	**Principle (arché)**	**Vision of the diffuse psyche**	**Connection with panpsychism**
6th century BC	Anaximander	Apeiron (infinite)	Everything comes from and returns to a limitless, eternal, and self-generating principle.	Conceptual basis for an eternal and animated cosmos.
6th century BC	Anaximenes	Air	Air is the vital breath that animates the universe and unites all things.	Precursor to the idea of the breath of life present everywhere.
6th–5th century BC	Pythagoreans	Number and cosmic harmony	The cosmos is an orderly structure where each part vibrates in harmony with the whole.	Introduction of mathematics as the language of energy and universal consciousness.
5th century BC	Heraclitus	Fire and Logos	Fire is living energy; Logos orders and permeates constant change.	Vision of a rational principle within matter.
5th century BC	Empedocles	Four elements + Love and Strife	Life arises from the interaction of opposing forces that animate the elements.	A universal dynamic of vital forces intertwined in matter.
5th century BC	Anaxagoras	Nous (Intellect)	A cosmic intelligence permeates and orders all reality.	Direct link between mind and matter, anticipating panpsychism.

Period	**Philosopher / School**	**Principle (arché)**	**Vision of the diffuse psyche**	**Connection with panpsychism**
4th century BC	Plato	Soul of the world	The cosmos is a living being with a rational soul that permeates all its parts.	Metaphysical foundation of a living and conscious universe.
4th century BC	Aristotle	Entelechy	Every being has an internal purpose and a vital force that guides its development.	Teleological connection between life, mind, and matter.
3rd–1st century BC	Stoicism	Universal pneuma	Vital and rational breath that permeates and connects all things.	Physical and unified vision of diffuse consciousness.
1st century AD	Neoplatonism	Emanations from the One	Reality is a gradient of life and consciousness emanating from the supreme principle.	Spiritual and philosophical synthesis of the living cosmos.
15th–16th centuries	Renaissance (Giordano Bruno)	Universal soul	Matter and spirit are inseparable; the cosmos is infinite and alive.	Transition to a modern conception of the diffuse mind.
17th century	Leibniz	Monads	Every material entity possesses perception and appetite, even if minimal.	Structured philosophical formulation of panpsychism.
20th–21st century	Contemporary philosophy	Consciousness as a	The mind is diffused throughout the	Historical synthesis with a

Period	Philosopher / School	Principle (arché)	Vision of the diffuse psyche	Connection with panpsychis m
		fundament al property	universe, present in every component of reality.	current philosophical and scientific approach.

Plato and the world of ideas

Plato imagined the cosmos not as a simple container of forms, but as a great living organism. In Timaeus, the dialogue that perhaps more than any other reveals his cosmology, the philosopher describes a benevolent "Demiurge" who shapes chaotic matter according to an eternal model: the world of Ideas. The universe is thus born as an imperfect copy of an immutable perfection, but not devoid of life. Indeed, Plato speaks of the "Soul of the World," a vital principle that binds heaven and earth, stars and men, in a single network.

The Soul of the World, says Plato, envelops reality like an invisible breath. It penetrates the perfect geometries of the heavens, the regular motions of the stars, the symmetry of the seasons. The philosopher does not see astronomy as a cold inventory of celestial bodies, but as proof of living order. Athens, in the 4th century BC, looked at the night sky as a fabric of intelligence, a place where Orion and the Milky Way were not just mythical figures, but points in an animated design.

In the story of Timaeus, the universe has a spherical body—a symbol of perfection—and a soul, which allows it perception and movement. This vision was echoed in the gardens of the Academy, where Plato taught surrounded by olive trees and the murmur of the Ilissus River. For his students, discussing stars and ideas meant getting closer to the very life of the cosmos. It was a philosophy that combined mathematics and poetry, reason and myth.

Plato places the Soul of the world at the center of a hierarchical system: it interprets eternal forms and distributes them in matter. Man, participating in this universal soul, can know the true order of things. It is not just a matter of knowing, but of recognizing that one's mind is part of a larger mind, like a note in the celestial symphony.

This vision influenced European culture for centuries. Renaissance figures such as Marsilio Ficino took up the idea that the cosmos possessed a soul, seeing it as a bridge between God and nature. In Dante's heavens, the angelic spheres move the planets like instruments in a divine orchestra. In the art of Piero della Francesca, geometry becomes an image of the secret breath of the universe.

Platonism, with its dream of an orderly and living cosmos, has left a profound legacy in the philosophy of panpsychism. If the world is animated, then consciousness does not belong only to humans, but vibrates in every part of the cosmic circle. Plato did not speak of quantum physics, but his intuition of a universal mind continues to dialogue today with those who seek to understand whether the universe itself can be considered a living subject.

The Platonic idea of a living cosmos in the ritual practices of the time.

The Platonic idea of a living cosmos, as expressed in the Timaeus, was not confined to philosophical speculation: it found correspondence in the religious and ritual sensibility of 4th-century BC Athens and, more generally, in classical Greece.

For Plato, the universe was not a dead collection of substances and forms, but an organism endowed with the Soul of the world, harmoniously connected to eternal Ideas. This vision inevitably influenced the way the Greeks related to the sky, nature, and the gods. In public rituals and domestic cults, the cosmic order described by philosophers was celebrated and recognized through gestures, offerings, and ceremonies that reflected a deep respect for universal vitality.

In religious festivals, such as the Panathenaea in Athens, the procession to the Acropolis was not only a civic act but also a tribute to the goddess Athena, considered the guardian of order and wisdom, elements that in Platonic thought are intertwined with the rational structure of the cosmos. The celebration of the solstice or the phases of the moon, following astronomical calendars, had a religious and at the same time 'cosmic' meaning: the regular motions of the stars were observed and interpreted as signs of a universal mind.

In more restricted and philosophical circles, such as the Academy, forms of contemplation were practiced that had an almost ritualistic function. Observing the sky, studying geometric proportions, or meditating on numerical relationships was not pure theory, but a way of communing with the ordering intelligence of the world. This approach is similar, in some ways, to 'theoria' understood as sacred 'vision': the act of knowing became a form of worship. Originally, in classical Greece, 'theoria' meant 'vision' or 'observation' and could refer to a journey to attend religious or ritual events, such as a pilgrimage to the Olympic Games or a sanctuary. This linked the word to an experience that was not only visual but also spiritual. In philosophy—especially with Plato and Aristotle—*theoria* became the disinterested contemplation of truth: an activity of the mind that goes beyond practical utility, aimed at grasping the order and harmony of the cosmos.

For Plato, *theoria* was closely intertwined with the vision of the Good and the Idea: an almost mystical act, because knowing the truth meant rising towards the divine.

Although critical of the literal myths of traditional religion, Plato recognized the value of the sacred dimension when it was linked to understanding the cosmos. In his dialogues, the celestial body is described as a perfect sphere because perfect geometry was considered a sign of divinity. Even the Demiurge of the Timaeus is, in fact, a philosophical articulation of the divine figure: a benevolent architect who shapes matter to reflect ideal perfection.

Oracles, such as that of Delphi, were part of this cultural context. While remaining linked to the Apollonian tradition, their role as interpreters of a hidden order could be seen, in Platonic terms, as contact with the universal mind. A response was not only a divine message, but a fragment of the total knowledge that the living cosmos held.

In short, the idea of the living cosmos influenced both large public rituals and more intimate and intellectual practices. It provided a metaphysical framework that gave meaning to the cults of the stars, city festivals, and philosophical contemplation, uniting religion and knowledge in a single vision: the universe as an animated organism, worthy of veneration precisely because it was alive and intelligent.

The Platonic cosmic vision in the Orphic-Pythagorean mystery cults.

The Platonic idea of a living cosmos, animated by a rational and ordering Soul of the world, did not remain merely a philosophical concept; it became intertwined with initiatory and mystical practices, especially where thought encountered the symbolism of the mystery cults of the Orphic-Pythagorean tradition.

As early as the 6th century BC, the Orphics conceived of the universe as traversed by a single vital principle. In their hymns, often recited during secret rituals, the cosmos was described as a being of which man possesses a divine spark. The doctrine of the transmigration of souls—also central to Pythagoras—found a logical and metaphysical structure in Platonism: if the Soul of the world permeates all things, every single individual soul is part of that cosmic life and can, through a process of purification, reunite with the whole.

In the Academy, philosophical discussion could have a function similar to that of initiation. Plato reformulated Orphic images and symbols in a rational key, but did not erase their sacred aspect. The creation myth in the Timaeus, with the Demiurge ordering matter according to mathematical proportions, echoed Pythagorean visions of number as the essence of universal harmony. In Pythagorean rituals, silence, meditation, and the study of proportions were not mere mental exercises: they were gestures of alignment with the rhythm of the living cosmos.

In the Orphic mysteries, the soul's journey after death involved passing through regions governed by precise cosmic laws. Some initiatory texts described paths of light and divine figures guiding the soul toward a "reunion" with universal principles. In this imagery, Plato's Soul of the World could be interpreted as the ultimate reality that welcomes the purified soul, led beyond sensory appearances.

Fourth-century BC Greece was fertile ground for the overlap between philosophy and religion. Orphic rites celebrated the human microcosm as a mirror of the divine macrocosm, while Platonic cosmology provided the "intellectual coordinates" of that mirror. Whether celebrating an initiatory ceremony or a discussion under the portico of the Academy, the goal remained the same: to understand—and live—the connection with the animating principle that moves all things.

Stoicism and universal pneuma. The vital breath of the cosmos.

If Plato had imagined the cosmos as a living organism with a Soul of the world, the Stoics—starting with Zeno of Citium in the third century BC—translated that intuition into an even more physical and immanent conception. For them, there was no separate world of Ideas: everything that lives and moves is part of a continuous substance, permeated by pneuma, the universal 'breath' or 'wind'.

For the Stoics, pneuma was not a purely metaphorical concept. It was a mixture of air and subtle fire, capable of animating matter and giving it form. This principle, both divine and natural, pervaded every being, from minerals to gods, linking everything to the whole. Cleanthes, Zeno's successor, speaks of this in the Hymns to Zeus, where the god is identified with the same life force that orders the cosmos, not distant but present in every fiber of reality.

In the Stoic schools of Athens, probably under the painted portico (*Stoa Poikile*), philosophy was taught as a discipline of life, but also as a science of the living universe. The study of natural laws was not distinct from morality, because understanding pneuma meant understanding one's place in a larger organism. The virtuous man was the one who lived 'according to nature', that is, in harmony with the function assigned to him by the universal breath.

The relationship with the sky, also central to Plato, became an element of daily participation for the Stoics. The observation of the cyclical nature of the sun and moon, the seasons, and atmospheric phenomena was not just scientific curiosity but recognition of the cosmic logos: the rationality intrinsic to the pneuma. Every act of knowledge and self-control was therefore a religious exercise, even if it did not involve a separate cult; it was an inner ritual to keep one's soul attuned to the universal breath.

This conception would profoundly influence Roman thought. Marcus Aurelius, in his 'Meditations', writes of himself as a 'part' of the whole, linked to a larger body by the same vital substance. What Plato had placed in an ideal world became, for the Stoics, a tangible and continuously perceptible principle.

In terms of historical continuity, Stoicism consolidated and made concrete the ancient dream of a living cosmos, offering a bridge to medieval and Renaissance conceptions of the universe as an organism. The idea of the universal pneuma was grafted onto ancient medical studies, Neoplatonic speculations, and alchemical doctrines, where the "breath of life" became the key to the transformation of matter.

From Stoic pneuma to the early developments of panpsychism.

The Stoic concept of pneuma—a vital breath composed of air and subtle fire, present everywhere and capable of organizing matter—can be seen as a conceptual ancestor of panpsychism. In fact, just as the latter theory holds that consciousness (or some rudimentary form of experience) is a fundamental and widespread property in the cosmos, Stoicism imagined a unified, immanent force that does not radically distinguish between matter and life.

From this perspective, the Stoics rejected the clear separation between spirit and body. Pneuma is both a physical and rational principle (logos), anticipating the panpsychist idea of a continuum between the material and mental dimensions.

For the Stoics, pneuma manifests itself with different levels of "tension" (tonos), from inorganic cohesion (minerals) to conscious life (human beings). This is reminiscent of some modern formulations of panpsychism that speak of "degrees" or "layers" of diffuse consciousness.

The notion of a universal life principle spread in Hellenistic medical texts and then in alchemical treatises, where matter was conceived as alive and in transformation. These traditions influenced Neoplatonism and, in the modern era, authors such as Giordano Bruno and Spinoza, who offered more philosophically structured versions of a mind diffused throughout nature.

The transition to modern panpsychism occurs when this universal "breath" is no longer interpreted as a specific physical substance, but as a fundamental property of being. This happened in the thinking of Leibniz, with his monads, and continues in contemporary debates on 'consciousness as a basic element', present in philosophers such as Galen Strawson and Philip Goff. Here, the historical root is maintained: the totality of the universe is alive or, at least, endowed with an inner dimension.

Aristotle and the vital form. Entelechy as the animating principle.

In the horizon of Greek thought, Aristotle remains a figure who marks both a break and a continuity with the previous world. Born in 384 BC in Stagira, a city on the Chalkidiki peninsula, the young Aristotle grew up on the margins of mainland Greece, far from Athens, but arrived there at the age of twenty to enter Plato's Academy. There he absorbed the spirit of Platonism, but soon realized that reality cannot be explained solely through the world of ideas. His gaze turned to the concreteness of living beings and their internal dynamics.

When Aristotle wrote *De* Anima, his treatise on the soul, he was already a mature thinker. He was interested in the psyche not as a separate entity, but as a principle that informs and enlivens matter. He thus introduced the concept of *entelechy*, a term derived from *en télos* échein, 'having the end within itself'. , Aristotle believed that every living being possesses within itself a principle of organization, a force that tends towards the realization of its full form. This force is not an external addition: it is intrinsic to matter, just as the energy of a flame belongs to fire itself.

In 4th-century BC Greece, nature was still populated by divine presences in the common perception. The fields, the sea, and the forests were considered to be inhabited by vital forces. Aristotle moves within this symbolic universe but makes a decisive shift: while recognizing the vitality of the world, he places the explanation of the natural order in a principle internal to the very structure of beings. The entelechy of a seed, for example, is what leads it, if conditions allow, to become a plant.

A famous Aristotelian description concerns the chicken egg. If it is fertilized and kept at a suitable temperature, life begins to manifest itself. Heat does not create life; rather, it serves to awaken the latent form that is "in potential" in the egg. In this perspective, the entire cosmos becomes a web of vital forms, each of which tends toward its own realization.

Aristotle spent years studying animals and plants on the island of Lesbos, in a setting reminiscent of an open-air laboratory. He observed fish in brackish lakes and rare species of shells along the Gulf of Kalloni. These empirical investigations reinforced his idea that life is not an accident, but the consequence of an active principle that permeates matter.

The notion of entelechy, while rooted in Aristotelian biology, soon took on a broader philosophical value. It became the cornerstone of a vision in which reality is teleological: every process is oriented toward an end. This thought would influence medieval philosophy, Renaissance debates, and modern metaphysical speculations, up to certain developments in contemporary panpsychism.

Ultimately, for Aristotle, life is the result of an internal form that acts silently but constantly, bringing every be ly towards its own fulfilment. It is a principle that, although formulated over two thousand years ago, continues to challenge science and philosophy on what we call consciousness and how it is rooted in the very fabric of existence.

How Aristotelian entelechy differs from the modern concept of 'purpose' or 'function' in biology.

The difference between Aristotelian entelechy and the modern concept of 'purpose' or 'function' in biology is profound, and concerns both the philosophical framework of reference and the way we talk about purpose in life.

Aristotle introduces entelechy in a metaphysical context. For him, nature is intrinsically teleological: every living being contains within itself, as a constitutive principle, the tendency to realize its full form. This tendency is not an external project, it is not the result of a design imposed by a distinct agent, but is the very reality of being. An oak seed 'tends' to become an oak tree because that is its essence, its potential form acting from within.

In modern biology, on the other hand, 'purpose' or 'function' are interpreted in evolutionary and mechanistic terms. An organ has a function—for example, the heart pumps blood—not because it is "destined" to do so by an internal principle that guides its existence, but because, in the course of evolution, that activity has contributed decisively to the survival of the organism and has been preserved by natural selection. The distinction can be summarized as follows:

In Aristotelian entelechy , purpose is immanent, linked to the ontological form of being. Every part of the organism, every vital process, is a step towards the fulfillment of what that being "is" by nature.

In the concept of purpose/function in modern biology, purpose is a descriptive, not an ontological, concept. We speak of 'function' to describe an effect that, in evolutionary history, has increased the fitness of the organism. There is no intrinsic purpose, but an explanation based on efficient causes and selection.

An example clarifies this better. For Aristotle, a bird's wings exist "to fly" not only because they enable flight, but because flight is part of the realization of the bird's "form." Function is inseparable from being.

For a modern biologist, however, wings are the result of anatomical variations selected over time, which have improved the ability to hunt or escape from predators. Flight is not an end in itself, but a useful evolutionary effect, maintained because it is beneficial for survival.

In fact, entelechy places us in a universe where matter is guided by internal principles towards complete forms. Current biology, on the other hand, places us in a universe where forms are the result of blind processes, which we can only retrospectively interpret in terms of 'function' or 'adaptation'.

The influence of the notion of entelechy on the thinking of medieval philosophers such as Thomas Aquinas.

Aristotle's notion of entelechy played a central role in the philosophical edifice of Thomas Aquinas, and more generally in medieval scholasticism, because it offered a language and conceptual structure suitable for uniting Greek philosophy and Christian theology.

When, in the 13th century, Aristotle's writings returned to Europe through translations from Arabic and Greek, Thomas encountered a way of thinking that spoke of internal purposes and substantial forms. Entelechy became for him the ideal tool for explaining how and why every creature acts with a view to fulfillment.

Thomas reinterpreted entelechy in light of Christian doctrine. For Aristotle, the intrinsic form of a living being leads it toward its natural perfection. Thomas retained this idea, but placed it in a theological context: every form, and therefore every entelechy, ultimately derives from God, who is pure act and the final cause of all reality. If for Aristotle the end is inscribed in the very nature of things, for Thomas this end is inscribed by an intelligent Creator.

In his *Summa* Theologiae, Thomas describes the rational soul as the substantial form of man, that is, as his specific entelechy. It informs bodily matter and gives it life, order, and direction.

This life has not only a natural purpose (to grow, think, act), but also a supernatural purpose: the attainment of eternal bliss. Here we see the fusion between Aristotelian teleologism and the Christian vision of salvation.

Even in non-human organisms, Thomas uses the logic of entelechy to explain their behavior. A plant "tends" to bloom, an animal "tends" to feed and reproduce, and this movement toward self-realization is, ultimately, a reflection of the great universal order willed by God.

From a cultural point of view, this synthesis produced a lasting change. The Aristotelian idea of immanent finality was preserved, but set in a framework in which the entire universe is an orderly organism, animated by a vital principle that is rational and divine. Many thinkers after Thomas—from Bonaventure to Suárez—inherited this perspective and used it to discuss the hierarchy of beings, cosmology, and even ethics.

In short, Thomas Aquinas took Aristotle's entelechy and transformed it into a conceptual bridge between natural philosophy and theology, making it a pillar of medieval thought.

Aristotle's entelechy and Taoism.

The comparison between Aristotelian entelechy and the Tao of Taoism is fascinating because it brings into dialogue two ancient and profound visions of reality, born in very different cultural and linguistic contexts.

Aristotle's entelechy arose within the framework of 4th-century BC Greek thought (although it has roots in earlier developments in Greek thought). In a philosophical tradition that sought causes and ordering principles of nature, Aristotle developed the idea in a highly rational and classificatory context, based on empirical observation and logical deduction.

The Tao (Taoism) developed in China in the 6th-4th centuries BC, rooted in the wisdom tradition of Laozi and Zhuangzi. The

Tao is the 'Way', the ineffable principle that generates and sustains all things, but which eludes conceptual definition.

Regarding the nature of the principle, we can say that entelechy is a principle intrinsic to every being, which leads it towards the realization of its full form. It is linked to the internal and specific structure of each entity. The purpose is specific: the oak seed tends to become an oak tree, man tends to develop his own rationality. Instead, the Tao is a universal and undifferentiated principle, from which every form emerges and to which everything returns. It is not aimed at the realization of a predetermined "type," but describes the continuous flow of transformation and cosmic harmony.

As far as purpose and movement are concerned, in entelechy, movement is oriented: from "potential" to "actual," with a point of perfection that represents the fulfillment of being.

In Taoism, movement is cyclical and spontaneous, without a definitive end. Perfection is not a static condition to be achieved, but harmony with the natural rhythm of the cosmos.

If we evaluate the relationship with human consciousness, Aristotle sees man as the most 'perfect' living form because he is endowed with reason, and his entelechy culminates in the development of rational and contemplativ l life. Taoism, on the other hand, does not classify life forms hierarchically. Human beings find fulfillment not in dominating or perfecting nature according to a model, but in 'not acting against' (wu wei) the universal flow.

There are also differences in language and method. Entelechy is a technical concept, located within a system of natural philosophy and metaphysics. It can be analyzed, defined, and deduced. In summary, one could say that entelechy looks from within each being, identifying a principle that drives it towards a specific realisation plan. The Tao looks from outside and inside at the same time, grasping the totality as a continuous flow, where there is no 'individual' plan separate from the whole.

Both see an order in nature—Aristotle reads it as finalistic organization, Taoism as spontaneous harmony—but they diverge radically on how this order operates and on the role of the human element.

The idea of 'vital form' or 'animating principle' in current philosophical and scientific debate.

There are several areas, both philosophical and scientific, in which the idea of 'vital form' or 'animating principle' (concepts close to Aristotelian entelechy) reemerges in new formulations. Although today's discussion often avoids traditional teleological language, certain fundamental questions remain the same: what makes an organism alive? What coordinates and unifies its parts? Where does consciousness come from? Here are some contemporary examples in which this idea resonates.

Philosophy of mind and consciousness.

First of all, there are the so-called *emergentist theories*: some philosophers and neuroscientists argue that consciousness emerges as a global property of the organism, with organizational characteristics that cannot be reduced to the sum of local physical processes. This is reminiscent of the unifying function of entelechy.

Giulio Tononi's *Integrated Information* Theory (IIT) proposes that consciousness arises from the degree of integration of information in a system. Although based on scientific methodologies, it attributes a primary role to the 'structure' of the system in determining the quality of experience, echoing the idea of an internal organizing form.

Theoretical biology and biosemiotics

In the organismic approach, some biologists, such as Denis Noble, argue that organisms should be understood as coordinated systems, in which 'top-down causality' influences cellular and genetic processes. This refers to the idea of a 'form' that guides the parts.

In biosemiotics, organisms are understood as semiotic systems that interpret signals and respond in an integrated manner. Here, the principle of life translates into the ability to "read" and "act" in a manner consistent with a biological context.

Theories of complexity and dynamic systems.

In complex models (e.g., in physics and systems biology), it is evident that new levels of order emerge spontaneously from interactions between many parts. Self-organizing processes such as morphogenesis are reminiscent of Aristotle's idea of "development toward a complete form," but interpreted as emerging from local rules.

Environmental ethics and ecological philosophy.

In thinkers such as Hans Jonas (*The Imperative of Responsibility*), life is defined as the presence of a "center of activity" that tends toward its own continuation. Jonas explicitly draws on the spirit of entelechy to establish a moral responsibility toward all living forms.

In ecological and holistic perspectives, the ecosystem is seen as a "diffuse organism" with its own vital coherence, analogous to an animating principle on a planetary scale.

Biotechnology and artificial intelligence.

The debate about when a highly complex system (e.g., a brain organoid or advanced AI) can be considered "alive" or "sentient" often implicitly recalls Aristotle's question: what is the "form" that makes a system more than a mere machine?

In essence, the idea of a vital principle has been transformed into concepts such as emergence, integration, and systemic organization, but it continues to play a decisive role in current conversations about life, consciousness, and natural order.

Below is a summary table comparing Aristotelian entelechy and these contemporary examples, to show similarities and differences at a glance.

Table 5 - Comparison between Aristotelian entelechy and some of its conceptual resonances in the contemporary world.

Field / Concept	Aristotelian Entelechy	Contemporary Example	Similarities	Differences
Philosophy of Mind	Internal principle that brings living beings from potentiality to actuality, fully realizing their form.	Emergentism and IIT (Integrated Information Theory): consciousness as a property emerging from an integrated system.	Both see an internal organization that unifies and directs the whole system.	Entelechy is teleological and qualitative; IIT is quantitative and based on information integration calculations.
Biology	The substantial form (e.g., vegetative, sensitive, rational soul) structures the organism and guides its development.	Organismic approach and biosemiotics: the organism as an integrated system with top-down causality and interpretative capacity.	Unitary idea of the organizing principle of the organism.	Today explained in terms of networks, feedback, and information, without reference to metaphysical "forms."
Complexity theory	Natural development oriented towards an intrinsic goal; gradual transition towards the	Self-organization and emergence of ordered structures in complex systems.	Both describe an order that emerges from the process itself.	Entelechy is oriented towards a predetermined end; complexity sees multiple outcomes that are not

Field / Concept	Aristotelian Entelechy	Contemporary Example	Similarities	Differences
	completed form.			fixed in terms of purpose.
Ethics	The intrinsic purpose of every living being suggests a good to be pursued.	Hans Jonas – principle of responsibility: moral duty to preserve life as something of intrinsic value.	Recognition of the organizational and teleological value of life.	Thomism and Aristotle root the end in nature/form; Jonas reconstructs it as an ethical imperative for the future.
Biotechnology and AI	'Form' defines the nature and vitality of an entity.	Debate on artificial life and AI sentience: is there a 'form' or organization that makes something r living or conscious?	Both seek the principle that distinguishes mere mechanism from living/sentient.	Entelechy assumes this principle as a given of nature; today, research is being conducted into whether it can be created artificially.

Stoicism and universal pneuma.

In Athens, in the painted porticoes of the Agora, Zeno of Citium taught an idea that sounded both revolutionary and ancient at the same time. The Cypriot philosopher, active since the end of the 4th century BC, saw the universe as a single body, held together by a vital breath: the pneuma. It was not an isolated concept. A few decades later, the Stoic Chrysippus of

Soli defined it as "*the force that penetrates everything and keeps it in harmony.*"

For the Stoics, pneuma was not ordinary air. It was a subtle substance, mixed with air and fire, endowed with rationality. It flowed through stones and trees, as it did through human blood. It animated the stars and dictated the rhythm of the tides. In this view, the universe was not a sum of inert things. It was a living, coherent organism, governed by an immanent intelligence.

Seneca, in the first century AD, writing from Rome to Lucilius, recalled that "*all things are united in one body, like the limbs of a single being.*" Without the pneuma, he said, the world would dissolve. This idea had a profound influence on Roman spirituality. Marcus Aurelius, emperor and Stoic, meditated that "*the world is one being, the substance is one, and the law is one, common to all.*"

The Stoic pneuma was also moral responsibility. If every creature shares the same divine breath, hurting another person or destroying nature means hurting oneself. This universal connection explained Stoic cosmopolitanism: the true citizen belongs to *the cosmos*, not just to his own city.

Stoic schools described pneuma in varying degrees. At the lowest level, it gave cohesion to inanimate matter. In plants, it manifested itself as the force of growth. In animals, it became sensitivity. In humans, it was reason. And in the entire cosmos, pneuma was God, not transcendent but present in every fiber of reality.

The Stoics did not imagine a separate God. Their God was nature. Not an external creator, but the intelligent fire that continually builds, destroys, and recreates everything. The great *ekpyrosis*, the periodic cosmic fire followed by the rebirth of the universe, was part of this life cycle. Pneuma survived every catastrophe, like indestructible energy.

The vision of pneuma united physics and metaphysics, ethics and theology. It brought together observations on the behavior of matter with a religious sense of life. This unity of knowledge influenced Christian thinkers such as Origen who, while modifying the doctrine, retained the idea of an omnipresent Spirit. In modern times, the concept of a diffuse life force influenced Leibniz and 19th-century vitalism.

Today, in discussions of panpsychism, the Stoic pneuma returns as a historical suggestion. The idea of distributed rationality, present in the very structure of the universe, goes hand in hand with contemporary physical and cosmological theories that speak of omnipresent information or consciousness as a fundamental property of reality. Ultimately, Zeno's vital breath continues to circulate, transformed but recognizable.

The role of Seneca and Marcus Aurelius in the development of the concept of universal pneuma.

Seneca and Marcus Aurelius played a crucial role in keeping alive and refining the Stoic idea of universal pneuma and universe, transforming it from an abstract philosophical concept into a practical guide for ethical living.

A philosopher, politician, and writer of the 1st century AD, Seneca interprets pneuma in a strongly ethical and cosmological key. In his writings, especially in his Letters to Lucilius, he describes the universe as a 'single body' of which every being is an integral part. This unity is made possible by pneuma, defined as the vital principle that permeates matter and ensures its cohesion. Seneca emphasizes that without this divine breath, the cosmos would disintegrate.

For him, pneuma is not only a concept of Stoic physics, but a moral foundation: since every creature participates in this common breath, damage inflicted on one part of the universe affects the whole. This is the basis of Stoic cosmopolitanism: the wise man must behave as a citizen of the world, respecting humans, animals, and nature as parts of himself. This view influences the way Seneca approaches issues such as justice, compassion, and moderation.

Two generations later, in the 2nd century AD, Marcus Aurelius brings the concept of pneuma to the heart of his Meditations. For the philosopher-emperor, pneuma is synonymous with common substance and universal law. In his reflections, he reiterates that 'the world is one being' and that everything that happens is part of a rational order that involves every element of reality, from the stars to humans.

Marcus Aurelius translates this doctrine into an attitude of acceptance and responsibility: if everything is permeated by a divine and rational breath, then even pain, loss, and death are natural and necessary events, to be faced with clarity and without rebellion. Awareness of pneuma leads him to insist on universal brotherhood, self-discipline, and a sense of duty towards the common good.

Seneca emphasizes the vital connection between the parts of the cosmos and the moral responsibility that comes with it. Marcus Aurelius highlights the rational dimension of the pneuma, integrating it into a philosophy of resilience and serenity.

Both transform the Stoic idea of universal breath into a principle of daily life, capable of guiding not only thought but also action.

The influences of Stoicism and the concept of universal pneuma on subsequent thought.

Stoicism, with its concept of universal pneuma, left a profound mark on the philosophical and religious traditions that developed in subsequent centuries, influencing both Neoplatonism and early Christianity.

The Stoic pneuma, understood as a vital and rational principle that pervades everything, found an echo in the Neoplatonic idea of a unified emanation from the supreme source, the One. Plotinus (3rd century AD) did not use the term pneuma in the technical Stoic sense, but conceived of the universe as a living organism springing from the divine mind (Nous) and the cosmic soul. The analogy with pneuma lies in its connecting function: for the Stoics, pneuma connects and structures the cosmos; for the Neoplatonists, the Soul of the World distributes life and order to all its parts.

In this transition, the material and physical element of pneuma was elevated to a purely metaphysical principle, but the idea of living unity and pervasive rationality remained.

In early Christianity, thinkers of the 2nd and 3rd centuries, such as Origen and Clement of Alexandria, were attentive to the philosophical doctrines of their time. The Stoic notion of pneuma as a rational and vital breath was integrated with the biblical conception of Pneuma (*Holy Spirit*), a divine presence that enlivens and guides the community and the universe.

In this cultural dialogue, the Stoic pneuma provided a model of immanence: God was seen not only as transcendent, but also as intimately present in every creature and natural phenomenon. While not adopting Stoic physics, some Christian authors retained the idea of a spirit that orders and sustains the cosmos, transforming it into a theological key.

In Neoplatonism, the universal bond became a hierarchical principle of emanations, but retained the vision of a living and orderly cosmos. In Christianity, vital unity merged with the conception of a personal God present in creation, paving the way for the theology of creation as the ongoing work of a life-giving Spirit.

In both cases, the Stoic universal pneuma acted as a conceptual bridge: it allowed reality to be thought of as a continuum between matter, life, and intellect, influencing both metaphysical philosophies and religious cosmologies.

Below is a synoptic table that clearly shows how the concept of pneuma was transformed as it passed from Stoicism to Neoplatonism and Christianity, with similarities and differences. The table highlights how the Stoic *pneuma*, albeit with different variations, found continuity and transformation in

Neoplatonic and Christian visions, maintaining the idea of a vital and ordering principle.

Table4- Transformation of the concept of pneuma with textual quotations from Seneca, Marcus Aurelius, Plotinus, and Origen.

Aspect	Stoicism (Seneca)	Stoicism (Marcus Aurelius)	Neoplatonism (Plotinus)	Early Christianity (Origen)
Definition of the principle	*"Totum quod vides, quo involvimur, unus est; membra sumus eius."* (*Letters to Lucilius*, 95) — The whole is a single body, an nimate by a common breath.	*"The world itself is a single living being, composed of substance and soul."* (*Meditations*, IV, 40) — A living being animated and held together by the rational spirit .	*"The soul of the world, coming from the Intellect, spreads life in everything."* (*Enneads*, IV, 4) — An immaterial principle that enlivens the whole.	*"The Spirit of God pervades the universe and penetrates every creature."* (*Commentary on the Gospel of John*, II, 6) — An omnipresent and life-giving spirit.
Cosmic function	Cohesion of matter and order of the elements, thanks to the *pneuma* permeating everything.	The cosmic *logos* directs and harmonizes every natural event and transformation.	The Soul of the World orders the cosmos and connects the sensible to the intelligible.	The Spirit creates and sustains life, and guides the history of salvation.
Rational dimension	*"Quod regit omnia, ratio est."* (*Letters*, 113) — Universal reason governs	Acceptance of destiny as an expression of rational order.	The Intellect (*Nous*) is the source of cosmic rationality.	The Holy Spirit enlightens the mind and guides it toward divine truth.

Aspect	**Stoicism (Seneca)**	**Stoicism (Marcus Aurelius)**	**Neoplatonism (Plotinus)**	**Early Christianity (Origen)**
	the cosmos.			
Relationship with man	Every man participates in *the* divine *pneuma*, establishing universal brotherhood.	The citizen of the world lives in harmony with nature and with other men.	The human soul is a fragment of the Universal Soul, capable of ascending to the One.	Every believer receives the Spirit as a gift and spiritual guide.
Ethical resonances	Respect for all forms of life, justice, and moderation.	Self-discipline, acceptance of destiny, social cooperation.	Inner purification and union with the divine source.	Universal love and care for the community as a spiritual body.

Classical and medieval philosophy.

To speak of panpsychism is to enter into a worldview in which everything is alive, endowed with some form of consciousness or sensitivity. The formula Deus sive natura, made famous by Spinoza many centuries later, has its roots in ancient times, scattered with philosophical insights that, from the Hellenistic age to medieval thought, sought to explain·the invisible link between mind and matter.

The journey is long and fascinating. It begins with Plotinus, who lived in the third century AD in Alexandria and then in Rome, who saw the universe as an organism animated by the living One. His image *of the World Soul* (a source of life that flows everywhere, from the most distant star to a grain of sand) profoundly influenced Neoplatonic philosophy and, consequently, part of Christian theology. It is said that his students described him as a pensive man, immersed in thought as if he were already in direct contact with the cosmic source he spoke of.

Neoplatonism itself supported the idea of a universal consciousness: a single principle that permeates every being. This view influenced the thinking of the Church Fathers, especially figures such as St. Augustine, who interpreted God's presence in creation as a living and rational energy. In this transposition, philosophy met theology, giving rise to a common language between metaphysics and faith.

In the medieval Islamic world, philosophers such as Al-Farabi and Avicenna reworked these insights in the light of the Aristotelian heritage and Greek translations that had reached Baghdad. Avicenna, a physician and thinker of the 10th-11th centuries, spoke of *an 'active* intelligence' that illuminates and orders the cosmos. For him, the stars were not simply inert masses, but animated entities, part of a living system. The astronomical observatories of Isfahan and Cordoba were not only places of calculation, but centers of contemplation of the living world.

With Thomas Aquinas, in the 13th century, the debate shifted to the Christian theological framework. Aquinas reinterpreted Aristotle to argue that matter, while lacking consciousness in the human sense, possesses an intrinsic potentiality animated by form. This "substantial *form*" is the principle of order and life. In the cloisters of Paris and Naples, Thomas discussed with teachers and students how nature and creation are intertwined in a single divine design.

At the same time, medieval mysticism, from Hildegard of Bingen to Meister Eckhart, explored spiritual experiences that saw God present in every creature and phenomenon. In Hildegard's visions, the world appeared as an immense living body, permeated by light and harmony. Eckhart, from the pulpit in Strasbourg, spoke of a soul that "breathes" together with the world, inviting us to perceive being as a continuum between spirit and matter.

This introduction prepares us to follow, in the following paragraphs, an ancient thread that unites Plotinus' living One, the cosmic hierarchies of Neoplatonism, the active intelligences of medieval Islam, Thomas' animated matter, and the mystical visions that permeate the Christian Middle Ages. It is the thread of panpsychism, a concept that spans centuries and cultures, and which today is finding new attention in debates on contemporary consciousness.

Plotinus and the "Living One." The cosmic soul as the origin of everything.

Plotinus was born in 204 AD, probably in Lycopolis, Egypt, a province then immersed in Hellenistic culture and Greco-Roman spirituality. He studied in Alexandria, a cosmopolitan city and crossroads for philosophers, doctors, and astronomers. Here he met Ammonius Saccas, an enigmatic teacher who introduced him to Platonic thought. Later, around 244, Plotinus moved to Rome, where he founded a school and became a central figure for the intellectual elite of the empire.

For Plotinus, the universe is not a collection of separate objects, but a great living organism. At the top is the One: the absolute principle, beyond matter and beyond definition. From the One emanates the *Nous* (Intellect), and from this is born *the Soul of the World.* It is this that brings the cosmos to life, binding every galaxy, plant, and human being in an invisible web. His idea appears in the 'Enneads', a collection of treatises compiled by his pupil Porphyry. In those texts, Plotinus writes: *'The whole universe is a single living being, endowed with a soul that pervades it'*.

His students describe a man detached from material goods, but not indifferent to life. Plotinus refused to have his face painted, arguing that the body is only a vehicle for the best part of oneself. He lived as if his mind were constantly turned towards that One he described as a "*pure source of light.*" Rome in the third century was a harsh and fragmented world, but in Plotinus' circle, the atmosphere was different: here, they discussed stellar souls, the harmony of the heavens, and how every creature is part of a living, intelligent design.

For Plotinus, the metaphor of the cosmic Soul is not a rhetorical ornament, but a philosophical reality. Everything—from the waves of the ocean to silent stones—is touched by this vital energy. The idea recalls certain Stoic intuitions about the pneuma ( ), the breath that embraces everything, but Plotinus takes it further. It is not just physical balance, it is widespread consciousness. The universe thinks, and it thinks in us.

This concept would influence late antique and medieval philosophy. St. Augustine, who read Plotinus in Latin translations, would find affinities between the Soul of the World and the presence of the Spirit of God in creation. In the schools of Athens and Constantinople, Neoplatonism would become a bridge between Greek philosophy and Christian theology, preparing the Middle Ages to see the cosmos as an animated body.

Plotinus died in 270 in Minturno, in southern Italy, far from the clamor of Rome. It is said that his last days were peaceful, devoted to contemplation. According to Porphyry, as his health declined, Plotinus said he wanted to "*unite the divine within us with the divine in the universe*." A simple phrase, but a perfect summary of a life spent looking beyond the boundaries of the visible.

Plotinus' idea of the One and monotheistic conceptions of God.

At first glance, Plotinus' One and the God of monotheistic traditions share a common trait: both are conceived as supreme principles of reality. However, their nature and their relationship with the world are profoundly different.

For Plotinus, the One is an absolute principle, beyond all form and determination. It is not a "personal" being with will, emotions, or intentions. It is pure unity, the source from which all things emanate. From the One comes the Nous (Intellect), and from this comes the Soul of the World. This flow is necessary, eternal, and impersonal: the One does not "decide" to create, but radiates reality just as the sun radiates light.

In monotheism, on the other hand, God is generally understood as a conscious and personal subject. In Judaism, Christianity, and Islam, God possesses will and acts in history. God "speaks," "commands," "loves," and "punishes." God is a relational entity, capable of establishing a dialogue with humanity and intervening in time.

Another crucial difference lies in the relationship with the world. For Plotinus, the One does not govern events like a sovereign, nor does it interfere with them. The cosmos is alive because it participates in the universal Soul, but the One remains beyond all change and is not altered by what happens. In monotheistic religions, on the other hand, God maintains a direct link with creation: he sustains it, guides it, and judges it.

Plotinus also places the One in a dimension of absolute transcendence even with respect to thought: the One is "*beyond being*" and "*beyond intellect.*" It does not think, because thought implies duality between subject and object. The monotheistic God, on the contrary, is conceived as supreme intellect and supreme consciousness—capable of knowing himself and all creation.

This distinction has profound implications. Plotinus' mystical universalism sees the return to the One as a process of internalization and dissolution of individuality. Monotheism, on the other hand, conceives of salvation as an encounter between two distinct subjects: the creator and the creature.

The intertwining of Neoplatonic currents and medieval Christian thought.

Augustine of Hippo came into contact with Neoplatonic thought at a crucial moment in his life. It was 384 AD, and the young rhetorician left Carthage for Milan, where he was to teach eloquence at the imperial court. There he read the Latin translations of Plotinus and Porphyry's Enneads, made by Mario

Vittorino. Those pages became a revelation for him: they offered a vision in which the world is not simply matter, but the manifestation of a higher principle.

Augustine found in Plotinus' One a key to understanding the transcendence of the Christian God. However, his God is not an impersonal principle: he is a conscious creator, capable of loving and intervening in history. If Plotinus' One emanates the cosmos out of necessity, Augustine's God creates it out of free choice. Herein lies the first fusion and the first tension.

Neoplatonism also offers him a language to describe spiritual ascent. Plotinus teaches that the soul, in order to return to the One, must detach itself from the sensible world and contemplate intelligible reality. Augustine translates this dynamic into Christian life: man must detach himself from earthly goods and turn toward God. Thus, conversion is not only moral but also ontological—a return to the true source of being.

In the Confessions, Augustine recounts a mystical experience he had with his mother Monica in Ostia. In that moment of contemplation, he says, "we *touched with our hearts the eternal Wisdom, which is above all things*." The image recalls Plotinus' passages on the ineffable contact with the One, but in Augustine this wisdom is identified with Christ. The bridge between Greek philosophy and Christian theology is completed: Neoplatonic language and concepts are transfigured by faith.

The Neoplatonic influence extends to his thinking on the divine presence in creation. Plotinus' Soul of the World becomes, in Augustine, the constant action of the Holy Spirit that enlivens all things. It is not a neutral principle, but a person of the Trinity. In this passage, universal consciousness is no longer an impersonal field, but the living breath of God in history and nature.

For medieval readers, Augustine is not only a theologian but also a translator of ancient visions. Thanks to him, Neoplatonism takes root in Christian thought, influencing monks, mystics, and scholastics for centuries. In Carolingian monasteries, in the cloisters of Cluny, works are studied in which the line separating the One and the personal God is thin, almost transparent—because the heart of the matter remains the same: the universe is animated by an intelligent presence, and man is a living part of it.

Influence of Plotinus' Cosmic Soul on subsequent theories of consciousness and nature?

For Plotinus, the Cosmic Soul is not a poetic metaphor, but the real and living principle that unites everything. It is the manifestation of the One within the sensible world, a field of vitality and intelligence that pervades the entire universe. In this perspective, consciousness is not an exclusive property of human beings: it is diffused in varying degrees throughout the cosmos, from plant life to the stars. This idea provided a conceptual basis for archaic forms of panpsychism, in which matter itself is animated by a mental principle.

In Christian thought, especially through Augustine, Plotinus' Cosmic Soul was reinterpreted as the presence of the Spirit of God in creation. Universal consciousness thus becomes divine will at work, capable of guiding nature and maintaining cosmic order. This influenced medieval philosophy, where nature is seen not only as creation, but as a living organism, ordered by a higher intelligence. Medieval mystics, from Meister Eckhart to William of Saint-Thierry, took up this fusion of cosmology and spirituality, speaking of a "breath" that animates everything.

In the medieval Islamic world, philosophers such as Al-Farabi and Avicenna adopted and elaborated on the Neoplatonic concept, integrating it with their metaphysics. The Soul of the World became the active intellect, the principle that infuses form and knowledge into matter. Consciousness, in this view, emerges because the structure of the universe is already permeated with intellect.

In later times, the Renaissance rediscovered Plotinus through Marsilio Ficino, who translated the Enneads into Latin. His idea of an "animated world" influenced the emergence of vitalist and pre-scientific visions, fueling the belief that every natural phenomenon is an expression of an intelligent order.

Elements of Plotinus' vision have also resurfaced in modern and contemporary philosophy. Spinoza, with his Deus sive Natura, proposes a unity between mind and matter that recalls the cosmic continuity of the One. Closer to us, currents of thought linked to quantum consciousness and contemporary panpsychism see in matter an intrinsic dimension of experience, a distant echo of Plotinus' idea that nothing is truly devoid of inner life.

In summary, Plotinus' Cosmic Soul laid the foundations for a conception of consciousness as a universal principle, intimately linked to the structure of nature. From a metaphysical intuition, it has become a common thread linking ancient philosophy, medieval theology, and modern scientific and philosophical debates.

The following table provides a comparative overview of Plotinus' panpsychism and contemporary theories on quantum consciousness, showing how these ideas have been transformed into the language of modern science.

Table5- Comparison between Plotinus' panpsychism and contemporary theories on quantum consciousness.

Aspect	**Plotinian panpsychism (Cosmic Soul)**	**Contemporary theories of quantum consciousness**
Basic principle	The universe is permeated by a Cosmic Soul, a vital and intelligent principle emanating from the One, present at every level of being.	Consciousness emerges from or is intrinsic to the quantum structure of reality; the quantum field, or quantum coherence, may be the "substrate" of the mind.
Origin of consciousn ess	Necessary emanation from the One: consciousness is universal and graded,	Result of quantum coherence processes in matter, especially in the

Aspect	Plotinian panpsychism (Cosmic Soul)	Contemporary theories of quantum consciousness
	more intense in intelligent entities, less so in material forms.	brain (e.g., Penrose and Hameroff's Orch-OR model), or a fundamental property of reality (physical panpsychism).
Relationship between mind and matter	Mind and matter are aspects of the same animated continuum; matter is "imbued" with intellect.	Mind and matter are connected by quantum phenomena; consciousness is linked to the configuration and behavior of subatomic particles.
Purpose/teleology	The return of the individual soul to the One through contemplation and union with the Cosmic Soul.	Some theories merely describe "how" consciousness emerges, without purpose; others hypothesize an evolutionary goal linked to universal information.
Nature of consciousness	It is a primary ontological quality of the universe, present everywhere; the cosmos is a living organism.	It is a physical or emergent property of the quantum system; there is ongoing debate as to whether it is primary or derivative.
Evidence and methodology	Philosophy and mystical intuition; contemplative experience as a tool for "recognizing" the Cosmic Soul.	Mathematical models and experimentation in neuroscience, quantum physics, and biology; search for quantum correlates in consciousness.
Vision of nature	Nature is sacred because it is part of the Cosmic Soul; every element participates in life and intellect.	Nature can be seen as a complex informational system; any diffuse consciousness is explained in physical terms.
Type of universal connection	Spiritual and ontological connection: everything is united because it is an emanation of the One and participates in the same soul.	Physical and informational connection: the quantum field connects distant systems through entanglement and coherence.

If the One is not a personal God, how does it interact with or influence the world according to Plotinus?

According to Plotinus, the One does not interact with the world in the sense of a personal God who decides, intervenes, or modifies individual events. The One is not an entity endowed with discursive will or emotions: it is the absolute and ineffable principle on which everything depends, but which remains beyond all form, thought, and direct relation; its "influence" on the world does not occur through action, but through emanation:

- From the One flows the *Nous* (Intellect) — an intelligible reality, the seat of all forms and archetypes.
- From the Nous springs *the World Soul* — the vital and intelligent principle that permeates the cosmos, orders it, and keeps it alive.

All beings, including humans, participate in this universal soul and are therefore, in some way, connected to the One.

The One is like a source of light that illuminates without "deciding" to do so: its nature is such that it inevitably radiates being and beauty, without loss of self. This continuous flow of emanation constitutes the connection between the One and the world, even without personal management or volitional intervention.

In other words, the One does not govern, but transcends and establishes; the world is alive and orderly because it participates in the Universal Soul, not because it is administered by an act of divine will.

Plotinus manages to reconcile the idea of the Soul of the World with the absolute transcendence of the One through a hierarchical and fluid vision of reality.

For him, the One is the supreme principle, beyond all form, thought, and difference. It does not act in the world as a personal God would, it does not make decisions and does not intervene in individual events: its presence manifests itself as a pure and silent source from which everything derives. It is precisely this overflowing ontological fullness that necessarily gives rise to the Nous – the Intellect – which contains the forms and archetypes of everything.

From the Nous proceeds the World Soul, the vital principle that permeates and orders the entire cosmos. It is like a universal breath: it enlivens every creature and maintains the cohesion of the universe, but it always does so while maintaining a link with the Nous and, indirectly, with the One. Its function is to make present, in a dynamic and immanent form, what in the One is silent transcendent power.

In this way, the One remains infinitely pure and distant, like light that does not become dirty as it illuminates. The World Soul is only the distant manifestation of that light, transmitted through successive passages that preserve its original transcendence. Thus, the cosmos lives and breathes thanks to the World Soul, without ever affecting the inviolable stillness of the One.

Neoplatonism and universal consciousness. Influences on Christian theology.

When Plotinus wrote the Enneads in the third century AD, the Mediterranean was a crossroads of cultures: Greek philosophies, Eastern religions, mystery cults, and nascent Christianity intertwined in the libraries of Alexandria and the porticoes of Rome. His vision of the One, a living reality from which everything emanates, was not just an abstract system. It was an attempt to explain how the entire cosmos could be thought of as an animated organism, in which every part is connected to a universal principle.

Neoplatonism, developed by his disciples such as Porphyry and Iamblichus, translated this thought into a complex cosmology. The Soul of the world, according to them, is the force that sustains matter and life. It is not a metaphor but a true spiritual entity. This idea soon entered into dialogue with Christianity, especially when Christian philosophers such as Origen or, later, Augustine of Hippo, began to question the relationship between God and creation.

Augustine, who was trained first in Neoplatonic philosophy and then in theology, saw in Plotinus' One an anticipation of the Christian God . However, he made a decisive shift: while for Plotinus the One necessarily spreads, for Augustine God creates voluntarily. It is a subtle but crucial change. The Neoplatonic universal consciousness, which permeates everything like light, becomes in the Christian vision a manifestation of divine will.

Mystical elements were also intertwined in this dialogue between two cultural worlds. In the Syriac monasteries of the fourth century, Neoplatonic texts translated into Syriac were studied, where the idea of the Soul of the world was interpreted as the immanent Holy Spirit. The Christian East, more sensitive to metaphysical nuances, welcomed the notion of a 'living' cosmos while maintaining the distinction between Creator and creation.

Neoplatonism also influenced Christian theology in the thought of Pseudo-Dionysius the Areopagite, a mysterious figure from the end of the 5th century. In his writings, the angelic hierarchies are presented as channels of spiritual energy coming from God and spreading to every level of existence. It is not difficult to recognize Plotinus' emanative structure behind this scheme.

In the Latin Middle Ages, the texts of Augustine and Dionysius arrived in universities thanks to the translations of John Scotus Eriugena in the 9th century. Thus, the Neoplatonic idea of a universal consciousness was reinterpreted as the "participation of being" of creatures in the uncreated light of God. In these classrooms, it was taught that nothing is without a connection to divine life. The stone, the plant, man, the angel—everything has a degree of consciousness insofar as it participates in the reality of God.

It is significant that even in places far from the cradle of Neoplatonism, such as the cathedral schools of Chartres, this cosmic vision was kept alive. In rose windows and frescoes, creation was depicted as a vibrant fabric of life, governed by an invisible divine mind. Neoplatonism did not therefore remain a philosophical doctrine: it became a spiritual lens through which Christianity viewed the world, seeing everywhere the trace of universal consciousness.

The Neoplatonic idea of emanation and the Christian doctrine of creation 'ex nihilo'.

The reconciliation between the Neoplatonic idea of emanation and the Christian doctrine of creation *'ex* nihilo' (from nothing) was one of the most delicate issues in late antique and medieval philosophy.

In short, the problem lay in the fact that for Plotinus and the Neoplatonists, the One necessarily 'emanates' the world, just as the sun emanates light: without an act of decision, without temporality. In the Christian view, however, God creates the cosmos through a free and intentional act, starting from nothing—no pre-existing matter, no cosmic necessity determining it.

Christian thinkers who absorbed Neoplatonism made a strategic reinterpretation of this concept.

Origen speaks of creative freedom. Origen (3rd century), educated in a context steeped in Platonic thought, replaced the necessity of emanation with an idea of voluntary *"procession"*: God, being infinite goodness, chooses to communicate himself by creating spirits and minds, but this choice is not obligatory. Emanation thus becomes an act of love.

Augustine cites the analogy of light. Augustine of Hippo, who had read Plotinus in Latin thanks to the translations of Mario Vittorino, retained the analogy of light spreading, but introduced a theological time: the light of creation begins to exist only when God makes it so. For Augustine, therefore, the world constantly 'depends' on God as emanation, but its beginning is the result of creation ex nihilo. Necessity is transformed into 'continuous dependence'.

Pseudo-Dionysius proposes the hierarchy of procession. In the texts of Pseudo-Dionysius the Areopagite (late 5th - early 6th century), the term "emanation" disappears, replaced by "procession." Procession is orderly, mediated by the heavenly orders, and starts from God through increasingly less luminous levels until it reaches matter. However, Dionysius always emphasizes that God is free and transcendent: his diffusion in creation is not an automatic motion, but a gift.

In the Latin Middle Ages, a compromise began to take hold. Authors such as John Scotus Eriugena and, later, Thomas Aquinas, found a point of balance. Thomas stated that every creature "participated" in the being of God, taking up the Neoplatonic idea of a universal ontological link. However, he insisted on creation ex nihilo, which occurs by divine will and not by natural necessity.

The flow of life that the Neoplatonists called emanation becomes, in Thomas, 'preservation in nothingness': God not only creates from nothing, but maintains everything in being.

In this way, Christianity preserved the mystical and cosmic dimension of Neoplatonism—the idea of a universe permeated by a living principle—without renouncing the theological foundation of free creation.

Creation 'ex nihilo' and quantum analogies.

The parallel between creation ex nihilo and some interpretations of the quantum vacuum is interesting. The Christian doctrine of creation ex nihilo states that God brings the cosmos into existence from absolute nothingness: no matter, no energy, no pre-existing physical laws. It is an ontological nothingness, that is, devoid of any reality, even potential.

The quantum vacuum, on the other hand, is not "nothing" in a metaphysical sense. In modern physics, the quantum vacuum is the lowest energy state of a field, but it is still "something": it has properties, fluctuations, and mathematical laws that are already operative. Virtual particles, pairs of particles and antiparticles, and observable phenomena such as the Casimir effect can emerge from this vacuum. This gives rise to analogies and differences.

One apparent similarity is that both concepts describe an origin from the absence of tangible material forms. Physics suggests that visible matter can emerge from a state that appears "empty" but is actually full of potential.

However, an essential difference would be that in theology, nothingness is devoid of structures and laws; in the quantum vacuum, physical laws are already present, and the possibilities for the emergence of matter are determined by these laws.

There is also a difference in causality: in creation ex nihilo, the cause is external to the world, that is, God, transcendent and free. In the quantum vacuum, phenomena emerge from fluctuations within the system itself, according to necessary and probabilistic rules.

Some contemporary philosophers, such as William Lane Craig and Robert J. Spitzer, have proposed that the quantum vacuum could be a "metaphorical" model to help us think about creation. The idea is that the modern mind, accustomed to scientific language, can imagine the creative act as the 'transformation' of a field without visible matter into a universe full of structures. However, the metaphor has an important limitation: the quantum vacuum already presupposes time, space and physical laws, while creation ex nihilo brings them into existence together with matter and energy.

This comparison between the quantum vacuum and creation ex nihilo has found its way into popular science, especially since the end of the 20th century with authors such as Stephen

Hawking and Lawrence Krauss, who have discussed the "*universe from nothing*" in physical terms. Christian theology, however, reminds us that the 'nothingness' referred to by science is not metaphysical 'nothingness'. Rather, it is an invisible but real 'something'.

The following table provides a comparative overview of "metaphysical nothingness" and "quantum vacuum" with examples that also link them to Neoplatonic thought.

This table clearly shows the ontological distance between the metaphysical "nothingness" of theology and the "vacuum" of physics, but it also highlights the points of contact that are useful for constructing cultural analogies. In medieval thought, the work of authors such as Augustine and Thomas made it possible to "translate" the lexicon of Neoplatonic emanation into a framework compatible with free creation, anticipating in some ways modernity, where scientific language seeks models of origin.

Table6- Relationships between the three concepts — Neoplatonic emanation, Christian creation ex nihilo, and quantum vacuum.

Aspect	**Neoplatonic emanation**	**Christian creation *ex nihilo***	**Quantum vacuum**
Concept ual origin	Philosophy of Plotinus (3rd century AD) and the Neoplatonic school.	Biblical and patristic theology, formalized above all in Augustine and Thomas.	20th-century theoretical physics, developed from quantum mechanics and field theory.
Causal principle	Necessary diffusion of the One, like the sun that emanates light without will.	Free and intentional act of God, transcendent and personal.	Fluctuations of quantum fields defined by physical laws.
Nature of "nothingn ess"/origi n	There is no such thing as "nothing": everything derives from an eternal	Absolute nothingness: total absence of matter, energy,	Physical state with minimal energy, but still "something" with

Aspect	Neoplatonic emanation	Christian creation *ex nihilo*	Quantum vacuum
	and pre-existing principle.	and laws before creation.	measurable properties.
Relationship with time	Act without temporal beginning: emanation is eternal.	Creation with a temporal beginning (even if God is outside of time).	Void defined within an already existing space-time.
Relationship with matter	Matter derived from the Soul of the world, which animates and orders it.	Matter and immaterial created simultaneously by God.	Particles emerge transiently from the void, according to physical probabilities.
Degree of con us consciousness/universality	The cosmos is permeated by an emanated universal consciousness.	Creation participates in the divine being but remains distinct from the Creator.	No implicit I consciousness; processes describable in physical-mathematical terms.
Medieval synthesis	Dionysius the Areopagite and Thomas reinterpret emanation as "voluntary procession."	Continuous creation (*conservatio in esse*) becomes a point of encounter with Neoplatonic language.	Analogy in the modern era: the quantum vacuum used as a popular metaphor, but with strong conceptual limitations.

Islamic thought towards the animated world.

In the heart of medieval Islam, the world was not seen as a collection of inert objects. It was perceived as a living organism, traversed by intelligent forces and an order that reflected the divine mind. Two key figures in this vision were Al-Farabi and Avicenna. These learned thinkers, educated on Greek works translated into Arabic, interpreted Aristotle and Plotinus in a way that combined philosophy, theology, and metaphysics.

Al-Farabi, born in Farab, Transoxiana (now southern Kazakhstan) around 872, worked in Baghdad, the beating heart of Islamic knowledge. He imagined the universe as a hierarchy of beings animated by a universal intellect. The One, the supreme principle, radiated levels of intelligence: each heaven had its own intellect. Matter was not dead, but received form and movement thanks to these celestial intelligences. Al-Farabi pushed Aristotle's idea further: he made the universe a perfect city, where every being played a specific role, like inhabitants coordinated by a great invisible ruler.

An anecdote tells that in discussions at the House of Wisdom in Baghdad, Al-Farabi amazed his listeners with his extraordinary ability to describe music as a reflection of cosmic harmony and th . For him, a musical interval was not just sound, but a sign of the internal proportion of being.

Avicenna (Ibn Sina), born near Bukhara in 980, inherited this vision and made it more subtle. For him, the universe was not only ordered by separate intelligences, but imbued with soul. The 'form' that everything possessed was a vital principle. The soul did not belong only to humans. Animals, plants, and even stones participated in a form of vitality, albeit to a lesser degree. Avicenna described the world as a flow of existence emanating from the First Intellect, where consciousness and matter were intertwined.

When Avicenna speaks of the 'soul of the sky', he seems to be reviving the ancient Neoplatonic concept of the soul of the world. But he translates it into the medical and physical language of his time. In his '*Kitab* al-Shifa', a monumental encyclopaedic system, one chapter describes how every celestial movement influences life on earth, like a breath passing from the firmaments to the roots of plants.

Baghdad, Bukhara, Isfahan: places where these ideas flourished were hubs of cultural exchange. Libraries housed Greek manuscripts translated into Arabic. Caravan routes carried not only goods but also philosophical concepts. In this environment, Islamic thought built bridges between Aristotle, Plotinus, and the Koranic revelation.

For Al-Farabi and Avicenna, consciousness was not a human privilege. It was a widespread quality. The universe breathed and thought, moved by a principle that united intelligence and love of order. This insight had a profound influence on Christian scholasticism, especially Thomism, but also on mystical currents that saw creation as alive.

For them, the idea of the 'animated world' was not just a poetic metaphor. It was a rigorous theory, built on Aristotelian logic and astronomical precision. But at its core remained a perception: the cosmos is a living being, and we are cells of a much larger body.

From the Islamic living cosmos to medieval panpsychism.

The vision of the animated cosmos developed by Al-Farabi and Avicenna did not remain confined to the libraries of Baghdad or the courtyards of Bukhara. In the 11th-13th centuries, while the Islamic world continued to translate and comment on Aristotle and the Neoplatonists, in Europe the university centers of Paris, Oxford, and Naples welcomed those same works, which arrived through Latin translations from Toledo and Sicily.

The idea that the entire universe was permeated by a vital principle found fertile ground in medieval European thought. In Christianity, this principle was interpreted in the light of theology: not an autonomous world soul, as in Neoplatonism, but a living order maintained by the constant will of God. However, the structural similarity was evident. Avicenna's '*active* intellect', which transmitted forms from the intelligible to the material world, was assimilated by scholastics such as Thomas Aquinas and Albertus Magnus as a useful element in explaining knowledge, causality and the perfection of the universe.

Al-Farabi had compared the universe to a perfect, hierarchically organized city. Avicenna had imagined every fragment of matter as a bearer of form and vitality. In medieval Christian synthesis, these images translated into a harmonious cosmos, where every creature—from the highest angel to the humblest speck of dust—participated in a vital current originating in God and organized by a hierarchy of causes.

This conceptual compatibility allowed for a kind of cultural "osmosis": Islamic panpsychism, while rooted in Aristotelian philosophy reinterpreted in a Neoplatonic key, offered medieval Christian theology a language and a mental framework for thinking about universal consciousness without falling into pantheism. The universe breathed with life, but that life came from the Creator, not from a neutral or impersonal principle.

In Benedictine monasteries and cathedral schools, the circulation of these texts transformed cosmology into a form of theological meditation. A monk could read Avicenna to understand the language of physics and the soul, and at the same time contemplate *'conservatio in* esse' as an extension of the divine breath.

Thus, the thought of Al-Farabi and Avicenna became a fundamental tile in the mosaic of medieval panpsychism: a bridge connecting Hellenistic cosmic mysticism, Islamic metaphysics, and Christian reflection on universal consciousness.

Thomas Aquinas and animated matter.

In 13th-century Paris, Thomas Aquinas walked the corridors of the Sorbonne with the calm of one who measures every thought. The Dominican friar, who had come from southern Italy, brought with him a box of conceptual tools: Aristotle translated from Greek into Latin, Arabic commentaries by Avicenna and Averroes, and the Christian legacy of Augustine. He did not use them separately. He intertwined them to give shape to a vision of the universe that, while remaining theologically orthodox, retained panpsychist undertones.

For Thomas, matter was not a mere lifeless extension. Every being in nature possessed a substantial form, an active principle that gave it identity and order. In an oak tree, the form was the vegetative soul; in a horse, the sensitive soul; in man, the rational soul. Matter did not act without form, and form did not exist without matter: they were an inseparable pair.

This conception derived from Aristotle, but Thomas grafted onto it the Christian idea of continuous creation. 'Form' was not a remnant of an eternal cosmos, as in Greek thought, but a constant gift from the Creator. Every form, even the humblest, participated in the divine being.

It is said that during the Parisian debates, a student asked if there was a vital principle in a stone. Thomas calmly replied: "*The stone does not have a soul like plants or animals, but its substantial form is what makes it a stone and not dust. Without that form, it could not exist.*" The answer was not a concession to pantheism, but a recognition that even what appears inert possesses a "determination" that comes from the act of creation.

In his *Summa* Theologiae, Thomas describes a cosmos organized as a ladder of forms, from the most perfect to the simplest. Each step was necessary. The lower forms prepared the ground for the higher ones, like the roots for the crown of a tree. This hierarchy was not cold. It was imbued with living order.

Islamic culture had given Thomas conceptual tools for thinking about the relationship between intelligence and matter. Avicenna's active intellect was transformed, in his hands, into a dynamic between God and the world: God imprints intelligibility on forms, and thus every entity becomes in some way "thinkable" and "speaking" to man.

The convent of San Domenico in Naples housed the manuscripts with which Thomas worked. Between the lines written by the copyists, there were often glosses linking Aristotelian physiology to everyday examples: the heat of the sun on the vineyards of Campania, the wind moving the waves in the harbor, the slow formation of marble in the quarries. These examples were not mere embellishments. They served to make visible the idea that matter, in all its forms, is part of a living design.

In Thomistic thought, animated matter did not mean that everything had consciousness. But it did mean that everything had an internal purpose, a "form" that led it towards its fulfilment. In this sense, Thomas embraced the ancient and Islamic sensibility for an articulated and vital universe, but subordinated it to God's creative freedom.

This is the core of his panpsychist nuance: a universe where nothing is inert, because even what does not think lives in the order of being, and from being receives strength and identity.

Medieval mysticism. Panpsychist visions in spiritual currents.

In the 13th and 14th centuries, in Europe, the vision of the cosmos as a living fabric left the university classrooms and found its way into monastic cells, the streets of merchant cities, and even the fields of peasants. It was no longer just a theoretical discourse. It became direct experience, often recounted in the language of mystical revelation.

Among the Beguines, women who lived in secular communities while devoting themselves to prayer, the Thomist idea of a creation permeated by divine life was translated into intimate and daring images. Hadewijch of Antwerp, in the 13th century, described the world as "*the one body of* God" where every being was a pulsating cell. She was not talking about heretical pantheism, but about an affective perception: God was not only the remote creator, he was the lifeblood that moved life in every blade of grass.

This sensitivity also found an echo in Franciscan circles. St. Francis of Assisi, already in the previous century, had sung of "brother sun" and "sister moon" with a sincerity that today we might interpret in a panpsychist key. For Francis, there was no clear boundary between matter and spirit. Every creature bore traces of divine love, and these traces were alive, present, palpable. His followers, especially the spiritual Franciscans, developed meditations in which the entire universe was seen as an animated community, a chorus of beings praising the Creator.

In these currents, the Aristotelian hierarchy of forms became a cosmic song. Angels, humans, animals, stones—all placed like different notes in the same h r melody. Some mystics went so far as to recount visions in which stones and flowers spoke, not with human words, but with "*languages of* light" or "*movements of* being" perceived in the depths of the soul.

In the convent of Helfta, Germany, Mechthild of Magdeburg recounts seeing creation as a large transparent circle. Every creature was connected to every other by threads of fire. Stones supported the weight of the universe, trees maintained its freshness, and humans carried the breath of the word. It was not poetry for its own sake. It was a way of interpreting theology: God permeates matter and directs it toward his glory.

This form of mystical panpsychism was not limited to ecstatic contemplation. It often guided concrete acts of care towards nature and animals. Some friars minor noted in their codices that moving an object, cutting a branch, or moving a stone required respect, because one was touching a fragment of God's living work.

The bridge between philosophy and mysticism was therefore solid. The idea, originating with Aristotle and refined by Thomas Aquinas, that every entity was animated by its substantial form, found an existential interpretation in the hearts of mystics: to live means to enter into relationship with a universal community of beings, each bearing a fragment of divine consciousness.

Hildegard of Bingen

Parallel to the birth of the new scientific cosmology and discussions on quantum consciousness, the Christian Middle Ages preserved a form of 'mystical panpsychism' that did not use this name but anticipated its central intuition: the world is alive, God is everywhere, matter is permeated by spirit. No figure expresses this vision better th r Hildegard of Bingen, a woman, Benedictine nun, mystic, and intellectual of the 12th century. Hildegard was born around 1098 in the Rhine region, in what is now western Germany. She lived in a Europe that was slowly emerging from the early Middle Ages, suspended between monasteries, crusades, and the birth of the first universities. Hildegard was many things at once: abbess, theologian, composer, healer, advisor to popes and emperors. Above all, she was a visionary, in the literal sense. From childhood, she said she saw lights, figures, and cosmic scenes that she interpreted as divine revelations. She built an organic system around her visions: a theology, a cosmology, a medicine, a music. It is in this intertwining that we glimpse a form of medieval panpsychism, where the divine manifests itself in every creature and phenomenon.

THE WORLD AS A LIVING BODY.

In her most famous texts, especially in Scivias (c. 1141–1151), *Liber vitae* meritorum, and *Liber divinorum* operum, Hildegard describes the cosmos as a great living organism, permeated by light, warmth, and harmony. In one of her cosmological visions, Hildegard sees the universe as an enormous oval figure, similar to an egg or a body, surrounded by divine fire. The celestial spheres, the winds, the stars, and the elements are not simply physical objects, but "members" of a great cosmic living being. God is not distant, not separate, but sustains and penetrates every part of this body. In a famous passage, Hildegard writes that man is a "microcosm" and the world is a "macrocosm." The human body mirrors the universe, and the universe resonates with the human body. When man sins, the cosmic order is affected; when man lives in harmony with God, nature also regains its balance. This vision is theological, not philosophical in the modern sense, but it conveys a strongly unified idea: there is a single life that flows, on different scales, in man, animals, plants, and elemen . This conception challenges the idea of pure matter as "res extensa," inert and separate from the Cartesian "res cogitans" of many centuries later. Hildegard's world does not recognize a clear dualism between matter and spirit. Matter is always already inhabited, illuminated, and animated by God.

LIGHT AS THE FABRIC OF REALITY.

A recurring theme in Hildegard's visions is light. Hildegard often speaks of a *'lux* vivens', a living light that is not only a metaphor for God, but is almost his way of manifesting himself in the world. In the miniatures accompanying the Scivias, created in the scriptorium of the monastery of Rupertsberg (near Bingen on the Rhine), the cosmos appears as an interweaving of circles, fires, stars, and luminous currents. The drawing is not a simple ornament: it is a visual cosmology. The universe is a translucent body, traversed by luminous energy that connects every level of existence, from angels to stones, from seasons to human thoughts. This insistence on light, understood as a dynamic principle, allows for a parallel—not an overlap—with the language of contemporary physics, where light and quantum fields are seen as the fundamental "fabric" of reality. Hildegard does not do physics, but she suggests something that modern science is partly rediscovering: the world is not a set of static blocks, but a field of processes, vibrations, and energy exchanges. In a panpsychist key, this '*lux* vivens' can be read as a sign of a diffuse interiority. Every creature is not only what it appears to be, but also a divine presence, a spark of meaning that participates in a greater horizon. Hildegard's universe is a universe that feels, because it is always exposed to the gaze and breath of God.

Contemporary panpsychism debates whether every physical entity—even an elementary particle—possesses a rudimentary form of experience or "proto-consciousness." Hildegard would never speak of the consciousness of an electron in . Her language remains theological and symbolic. However, her fundamental intuition is close: nothing is a pure "object."

Everything is included in a horizon of life, intention, and meaning. For Hildegard, God is present in every creature in different ways. In humans, God manifests as reason, freedom, and moral responsibility. In animals, God appears as instinct, movement, and life force. In plants, God shows itself as growth, fertility, and seasonal rhythm. Even minerals and elements contain an energy that has divine origins. The German nun uses a key term, "*viriditas*," "greenness," the vegetative force that makes the world fertile and alive. *Viriditas* is much more than a poetic image: it is the tangible trace of divine life in nature. When the fields are lush, when health flourishes in the body, when music elevates the soul, Hildegard sees this "green tension" at work, uniting spirit and matter. From this perspective, creation is not a machine, but a choir. Every creature sings its part. Man, whether male or female, has the task of harmonizing with this choir, not dominating it. Here an interesting connection opens up with current ecological discussions and with forms of panpsychism that see in the recognition of a diffuse "interiority" an ethical basis for respect for the environment.

VISIONS, ECSTASY, BODY.

Hildegard's mysticism is not disembodied. The nun experiences her visions in her body, often accompanied by physical discomfort, fever, and collapse. She herself recounts that she hesitated for a long time before writing down her visions, until an inner voice—which she identifies as a divine command—prompted her to obey. The dynamic is significant: the vision is not an abstraction of the mind, but an experience that involves flesh, breath, sight. The universe that Hildegard sees is a body, and that same vision passes through her body. A h l resonance is created between microcosm and macrocosm: the human body becomes the point where the sensible and spiritual universes meet and translate into each other. From a panpsychist perspective, one could say that consciousness—understood as subjective experience—is never entirely enclosed within the skull. For Hildegard, human consciousness is always in relation to and in exchange with the cosmos. Her visions show this in a dramatized form: the soul is not a spectator, it is a participant, a node in a living network that permeates everything. This insistence on corporeality recalls, centuries later, some insights of contemporary neurophenomenology, which speaks of embodied consciousness, situated, dependent on the environment and interactions. While modern panpsychism debates whether consciousness emerges from the brain or is a fundamental property of reality, Rhineland mysticism experiences consciousness as something that opens up to the universe, not as an internal archive.

A MEDIEVAL LABORATORY OF 'DEUS SIVE NATURA'.

To understand Hildegard, it is useful to imagine her surroundings. The monasteries of Disibodenberg and then Rupertsberg, overlooking the hills that dominate the confluence of the Rhine and the Nahe, were immersed in a landscape of vineyards, woods, mists, and waters. In that landscape, Hildegard wrote, prayed, observed the seasons, treated the sick with local herbs, and listened to the liturgical chants that resounded in the stone naves.

This context is not just scenery. It is part of her cosmology. The river, flowing slowly but inexorably, becomes a pattern for thinking about the flow of time and grace. The cycles of the vines and fields illustrate the logic of viriditas. The winds rising from the valley visualize the idea of a cosmic breath connecting heaven and earth. In the 12th century, new translations of Aristotle, Arabic texts, and Jewish commentaries began to circulate in Western Europe. In the cathedral schools of Chartres, Paris, and Laon, a natural philosophy was developed that sought to understand the order of the world through rational means ( ). At the same time, in the relative quiet of the Rhine, Hildegard proposed another form of knowledge: not logical deduction, but visionary intuition. Yet there are bridges between these two paths—scholastic and mystical. Both reject the idea of a chaotic cosmos. Both seek laws, forms, harmonies. It is significant that it was precisely from Chartres and other schools linked to Christian Platonism that ideas of a harmonious, geometric universe emerged, which, centuries later, would also enter the lexicon of Kepler and the first modern scientists. From this point of view, Hildegard's mysticism is 'another path' towards what, in Spinozian language, we might call *Deus sive natura*. Hildegard never identifies God and nature in a philosophical way, because she remains within Christian orthodoxy. For her, God transcends the world. But at the same time, God is so intimately present in the world that it becomes, in fact, a great living symbol of itself. Nature is not God, but it is so imbued with God that it cannot be reduced to a mere thing. Contemporary panpsychism, when seeking a non-theological language, often ends up talking about "mental properties" diffused in matter. If Hildegard could hear this debate, she would probably smile and respond with an image: a tree, a spring, a stone. She would say that everything is a sign, and that the real problem is not whether stones "think," but whether men and women are still capable of hearing the silent song of creation.

FROM MEDIEVAL VISIONS TO QUANTUM CONSCIOUSNESS.

What is the link between this visionary universe and current discussions on quantum consciousness and panpsychism? The obvious risk is forced overlap. Hildegard's cosmology is symbolic and theological, not physical. Quantum physics, in turn, does not speak of angels or virtues. However, there are some resonances that are worth exploring.

Unity of reality.

Hildegard insists on a deeply interconnected universe, where nothing exists in isolation. Current holistic interpretations of quantum mechanics—from entanglement to field theories—describe a world in which the physical states of distant systems can be correlated in ways that defy classical intuition. In both cases, the idea of "independent pieces" loses strength.

Diffuse interiority.

Modern panpsychism, especially in forms inspired by "*Integrated Information* Theory" or quantum models of the mind, hypothesizes that consciousness is linked to information structures or fundamental quantum states. Hildegard does not speak of information or qubits, but she is convinced that every creature possesses an "inner dignity" because it is in direct relation to God. Human consciousness, in this framework, is not an island; it is the point at which the universe becomes aware of itself in a reflexive way.

Symbolism of light.

Light is at the center of both Hildegard's mystical language and modern physics, which sees electromagnetic radiation and photons as key elements in the structure of reality. For the Rhineland mystic, light is divine presence; for science, it is a physical phenomenon. Yet both traditions use light to say that the world is dynamic, that something passes through it and connects it. Some contemporary authors, in search of a common lexicon between spirituality and science, have turned to figures such as Hildegard to show that the idea of a living cosmos did not originate with quantum physics, but preceded it by many centuries. It is not a question of retrospectively reading 'quantum' into her visions, but of recognizing a continuity of questions: what does it mean to say that the world is alive? Where does matter end and the mind begin? Is there really a clear boundary?

THE LEGACY AND RELEVANCE OF A PANPSYCHIST MYSTIC.

Today, Hildegard of Bingen is venerated as a saint and doctor of the Church, but she is also a figure of interest to musicologists, historians of science, ecologists, feminists, and philosophers of the mind. Her musical compositions—written for the female monastic community she led—are among the most original of the Middle Ages: extended melodies, bold intervals, texts that celebrate creation as a vibrant fabric of light and sound. Hildegard's music is perhaps the most direct form of her mystical panpsychism: each melody seems to want to give voice to the world. Not only does the human soul sing to God, but the entire creation sings in her and through her. In these compositions, the universe is no longer just a living body, but also a harmonious polyphony.

In the contemporary debate on consciousness and the universe, the figure of Hildegard acts as an alternative memory. She reminds us that the idea of a mental or spiritual cosmos does not necessarily arise from a rejection of rationality, but from a different experience of the world. For the Rhineland nun, reality manifests itself in a way that is at once sensory, symbolic, and spiritual. The mind cannot isolate one of these layers without losing something essential. While the philosophy of mind debates whether consciousness emerges from neurons or is a fundamental aspect of the universe, Hildegard's mysticism offers another image: consciousness as participation. Human beings, whether women or men, are conscious not because they possess a separate mental substance, but because they are placed in a web of relationships with God, with others, and with creation. Contemporary panpsychism seeks to say something similar in secular language: the individual mind is not a monolith, it is a node in a larger web. In this sense, the visions of the mystic of Bingen do not belong only to the Middle Ages. They continue to challenge us every time we try to think of consciousness beyond the brain, nature beyond inert matter, God beyond the categories of rigid theism. Her world, seen as an immense living body permeated by light and harmony, remains one of the most powerful pre-modern s of what we now call Deus sive natura and panpsychism.

From the late Middle Ages to the Renaissance. Panpsychism, Ficino, Bruno, and the echo of "Deus sive Natura."

At the end of the 15th century, Europe was a house in turmoil. Universities, still anchored in scholastic disputes, coexisted with noble courts where astrology, alchemy, and philosophy mingled. The direct rediscovery of Greek texts thanks to Byzantine refugees after the fall of Constantinople opened up a horizon that spoke with an ancient but fresh voice.

In this context, Marsilio Ficino—translator of Plato and director of the famous Florentine academy founded by Cosimo de' Medici—took up Neoplatonic ideas and gave them a Christian-humanist tone. For Ficino, the universe was a living organism, traversed by a world soul (anima mundi) that united heaven and earth. This was not a vague concept. The soul of the world acted as a vehicle of influence: planets, stars, plants, and humans exchanged energies and meanings through subtle threads. In his Florentine salon, Ficino discussed how musical harmony could attune the human soul to this universal current, quoting Pythagoras and Plotinus with equal ease.

The bridge to the Middle Ages was evident. The idea of substantial forms, which Thomas Aquinas had placed at the center of his theological cosmology, became in Ficino a living fabric in which each form participated in a greater consciousness. The hierarchy of entities was transformed into a dynamic set of relationships, where the human spirit could ascend the steps of being to the divine.

A few decades later, Giordano Bruno, a restless wanderer and ardent mind, would take this vision to a radical level never seen before. For Bruno, there was no 'outside' of God' . The Latin phrase Deus sive Natura, later taken up by Spinoza but already insinuated in Bruno's thought, meant that nature itself was the manifestation of the divine. Every point in space, every fragment of matter, every "atom" was animated by a vital and infinite principle. Not only the starry sky, but the roots of a tree or a grain of sand held a reflection of the Infinite.

Bruno brought these ideas to the squares and academies of France, England, and Germany with fiery, almost theatrical language. It is said that during a lecture in Paris in 1585, he compared the universe to "*a living animal that breathes in every part*." This was not an isolated poetic image. It was a philosophical statement and, in the context of the time, a dangerous one: dissolving the boundary between Creator and creation paved the way for accusations of heresy.

The panpsychist Renaissance, in the hands of Ficino and Bruno, brought together three cultural strands:

- the Neoplatonic and medieval tradition of the soul of the world;
- Aristotelian cosmology filtered through scholasticism;
- the nascent scientific and naturalistic sensibility, which viewed nature as an open book.

This synthesis, in the climate of intolerance and curiosity of the 16th century, produced a thought that was at once mystical, philosophical, and proto-scientific. In it, "Deus sive Natura" was not an abstract formula, but a vision of reality: living in a universe that thinks, feels, and creates, where consciousness is not a human privilege, but a widespread and unfathomable property.

Giordano Bruno and Baruch Spinoza.

Giordano Bruno and Baruch Spinoza, although separated by a century of history and very different cultural contexts, seem to dialogue ideally through their visions of the divine and nature. Both reject the image of a distant God, an external, immanent God, and instead propose a constant, intimate, immanent presence: God is not "above" reality but is reality itself, in its fullness and infinity.

In Bruno, this image takes the form of a boundless universe, scattered with infinite worlds, each imbued with a vital spark. For him, nature is alive in all its parts; everything, no matter how small, contains a fragment of divine energy. It is a vibrant, almost poetic vision that breathes in infinite space and feeds on hermetic and vitalistic suggestions.

Spinoza, on the other hand, translates this intuition into a rigorous and geometric language. In his philosophy, God and Nature are the same "infinite substance," unique and necessary, which expresses itself through infinite attributes and modalities. If Bruno sees every fragment of the universe as a center of life, Spinoza grasps its eternal and immutable structure, without attributing individual consciousness to each element, but recognizing that everything is a manifestation of the same reality.

Both criticize institutional religions, dismantling their claims to absolute truth. Bruno, with a more polemical and visionary approach, considers them human creations linked to historical circumstances; Spinoza, with analytical calm, reduces religion to its moral function and interprets sacred texts with critical tools.

Thus, if Bruno is a precursor of a vital and cosmic pantheism, Spinoza is its rational systematizer: one speaks the language of fire and stellar infinity, the other the clear language of necessity and logical order. Two different paths that, in the final analysis, converge in a common idea: the divine is not elsewhere, but in the very fabric of existence.

The Renaissance. From the soul of the world to 'Deus sive Natura'.

When Marsilio Ficino translated Plato and Plotinus into Latin for the Medici circle in the 15th century, the echo of medieval visions of the living cosmos found a new theoretical framework. The Franciscan idea of universal brotherhood and Beguine mysticism were grafted onto a system where the soul of the world was a real and operative principle.

Ficino did not speak of God as merely a transcendent creator. He described him as a force diffused in everything, a light that passes through matter and orders it. He believed that music could act as an instrument to "tune" the individual soul with the universal one. It was not an aesthetic exercise, but a cognitive and almost therapeutic act. In his study in Careggi, among Greek manuscripts and armillary spheres, the Florentine philosopher saw in the cosmos a great hierarchical harmony in which even stones were immersed in a subtle life.

Then came Giordano Bruno. Born in Nola in 1548, educated in the Dominican convent and the squares of Europe, Bruno took the idea of the animated universe to its extreme limit. For him, there was no center and no periphery: space was infinite, populated by countless worlds, and the same divine energy acted in every grain of matter. He did not make a clear distinction between creation and creator. It was here that the expression "Deus sive Natura" took on an almost programmatic significance, anticipating Spinoza's formula: God is Nature, and Nature is God.

This vision was more than a philosophical abstraction. For Bruno, it meant that the divine manifests itself everywhere, not only in a church or in a sacrament. Anyone who observes a plant or a star is already performing an act of sacred contemplation. It was a conception that amplified medieval mysticism and freed it from the margins of devotion to place it at the heart of a boundless cosmology.

The Renaissance was therefore a moment of dangerous and fascinating fusion: the contemplative tenderness of Francis of Assisi met with Bruno's bold infinity; the language of mystical love of a Hadewijch was translated into the categories of the universal soul and Platonic h l correspondences. From this union arose a thought that dared to say that the entire universe is not only inhabited by the divine, but is the divine itself in action.

From "Deus sive Natura" to contemporary theories of consciousness.

In modern thought, panpsychism has undergone a sort of silent metamorphosis. We no longer speak, as in the fifteenth century, of the "soul of the world" in a mystical-astrological sense, but we recover the same intuition: consciousness or, more generally, a form of elementary experience, is intrinsic to all material reality.

Philosophers such as Leibniz in the 17th century, with his theory of "monads," indirectly continued along a path opened up by Bruno: every entity is a center of perception, a complete microcosm. In the 20th century, thinkers such as Alfred North Whitehead formulated a '*process* philosophy' in which even an electron is an experiential event, connected to all others. These concepts are not just speculative curiosities: today they dialogue with neuroscience, physics and information theory.

The contemporary panpsychist hypothesis questions a problem known as *the 'hard problem of* consciousness', formulated by David Chalmers in the 1990s. If consciousness cannot miraculously emerge from nothing, perhaps it is already inscribed in the very structure of the universe. Here, the reference to the Renaissance vision is evident: the terms change, but the idea remains the same—thinking matter is not an exception, it is the norm.

In recent decades, a more daring theory has also emerged: quantum consciousness. Physicists and theorists such as Roger Penrose and Stuart Hameroff have proposed models according to which microtubules in brain cells could operate quantum processes capable of "hosting" states of consciousness. Not all scientists accept this view, but its appeal lies precisely in its echoes of Bruno's ideas: if consciousness is rooted in quantum phenomena, then it is not confined to humans, but present in varying degrees in everything that participates in the same physical laws.

From this perspective, today's 'Deus sive Natura' is no longer necessarily the theological God. It could be the network of quantum correlations that permeates every particle, or the totality of information/consciousness that structures reality. But the substance remains the same: the universe is alive, intelligent in its most intimate fabric, and our mind is just one of its countless expressions.

VI° Spinoza and "Deus sive Natura"

When Baruch Spinoza wrote his *Ethica* in the second half of the 17th century, he was living in a Europe riven by religious conflicts and political tensions. Yet, in a modest room in Amsterdam, a man cast out by his own Jewish community was developing an idea that was destined to remain one of the most radical in modern philosophy: Deus sive Natura. God, or Nature. Not two substances, but a single infinite reality.

Spinoza argued that there is only one substance, eternal and necessary. Everything that exists is a manifestation of it: humans, stars, a grain of sand. There are no ontological hierarchies: there is no "afterlife" distinct from the "material world." In this rigorous monism, God is not an external legislator but the totality of being itself. For Spinoza, saying "God" or saying "Nature" is saying the same thing.

One of the philosopher's most fascinating insights is the parallel conception of mind and body. They are not two separate substances, as Descartes believed, but two ways of grasping the same reality. Thought and extension are different attributes of the same unique substance. The body "thinks" in its own way and the mind "acts" through the body. Spinoza thus anticipates a non-dualistic vision that many neuroscientists today find in the concept of the integrated "body-mind."

In this perspective, every entity participates to some extent in universal consciousness. A stone does not 'think' like a man, but expresses, according to its order and degree, the same common reality. The panpsychist echo is evident: consciousness does not arise magically at a point in evolution, but is a fundamental feature of reality. This leads Spinoza to a rational ethic: acting according to reason means conforming to the necessary order of Nature, without illusions of separation.

Curiously, Spinoza's 'God-Nature' found sympathizers in scientists such as Albert Einstein. Einstein declared that he believed *'in Spinoza's God, who reveals himself in the harmony of what exists, not in a God who takes interest in the fate and actions of human beings'*. The idea of a unified universe, governed by necessary laws, has a family resemblance to modern physics, from the symmetries of relativity to the holistic visions emerging in contemporary cosmology.

Leibniz, while admiring the logical power of the Ethica, rejected its identification of God and Nature. He preferred a universe composed of individual, irreducible, and perceptive "monads." Schelling, in German Romanticism, on the other hand, saw Spinoza as a precursor of the philosophy of nature and attempted to reconcile Spinozian monism with idealism. In any case, the formula Deus sive Natura remained a point of reference, capable of provoking and inspiring even today. In reality, as already mentioned, Spinoza is effectively a pantheist, even though he is usually associated with panpsychist thought.

The Spinozian system. Substantial monism.

Baruch Spinoza was living in a modest house in Rijnsburg when he began to write down the ideas that would make him one of the most controversial philosophers of the 17th century. It was 1661, and Europe was slowly emerging from the shadow of the Thirty Years' War. Religious conflicts had prompted many thinkers to seek a universal truth beyond dogmatic disputes. Spinoza chose the most radical path: to assert that everything that exists is one and the same substance.

For him, God was not an anthropomorphic ruler observing the world from outside. God was the world itself, the totality of being. "Deus sive Natura": God, or Nature. This formula, which appears in the "Ethics" published posthumously in 1677, was not simply metaphysical poetry. It was the logical description of a universe without ontological separations. There is a single infinite, necessary, and self-sufficient reality.

Spinoza called this substance "*cause of itself.*" No creature generated it, no external force can destroy it. Every particular thing—a mountain, a cat, a human mind—is a "modification" of that substance. This idea broke down the great Cartesian dichotomy between mind and matter. In Spinoza's view, consciousness and matter are not opposing realms, but different expressions of the same original source.

In an imaginary dialogue between Spinoza and Descartes, set in the damp streets of Amsterdam, one might hear the Portuguese-Dutch philosopher say, "*René, you divide what is indivisible.*" Descartes would reply by talking about the clarity of thought as opposed to sensory confusion. Spinoza, on the other hand, saw in substantial unity a more parsimonious and elegant explanation of the universe.

Spinoza's monism was not, however, an isolated idea. The Renaissance had already glimpsed unitary conceptions of the cosmos, with figures such as Giordano Bruno. However, Spinoza gave this intuition a geometric and rigorous structure. Each proposition *of the Ethica* follows from axioms and definitions, as in Euclid. Philosophy becomes almost a mathematical edifice.

Culturally, Spinozian monism influenced thinkers and scientists far beyond his time. Three centuries later, Einstein spoke of 'Spinoza's God' as a harmonious order of the universe. Goethe read *the Ethics* and drew inspiration from it for his 'religion of Nature'. The idea of a single substance continues to inspire theoretical physicists and cosmology studies today.

The price Spinoza paid was high. In 1656, the Jewish community of Amsterdam expelled him with the *cherem*, the anathema. His monism was seen as a threat to the idea of a personal and providential God. The religious authorities—and most of the citizens—considered his identification of God with the totality of material and mental reality dangerous. Yet it is precisely this identification that is now recognized as one of the roots of modern panpsychist thought: consciousness and life are part of a whole that has no ontological boundaries.

The uniqueness of Spinoza's monism.

Spinoza's monism differs markedly from other forms of philosophical monism both in its rigorous conceptual structure and in its metaphysical and ethical implications.

Spinoza asserts that there is only one infinite and necessary substance, which he calls either God or Nature. Everything that exists—bodies, minds, events, thoughts—is an r modification of this same substance. There is no ontological separation between the mental and the material: they are two attributes of the same reality, perceived in different ways. This radical rejection of Cartesian dualism is the crux of his originality.

If we compare it with other forms of monism, important differences emerge. First, let us compare it with Leibniz's monism. For Leibniz, the universe consists of an infinity of simple substances called monads. Each monad is a closed and autonomous center of perception that does not interact causally with others but follows a "pre-established harmony" willed by God. Here, the unity of reality is not given by a single substance but by a divine concertation of separate entities. Spinoza, on the other hand, rejects the idea of multiple substances: what we call "entities" are only modifications of a single substance.

In idealistic monism, Hegel conceives of reality as the dialectical development of the Absolute Spirit. Unity manifests itself through a history of contradictions and overcoming, where consciousness progressively evolves towards total self-understanding. Spinoza, on the contrary, sees substance as eternal and immutable, devoid of temporal development. Modifications follow necessary but not teleological laws; substance does not "become" something, it is already everything.

Modern materialist monism, in some contemporary forms, reduces everything to matter and its interactions. This perspective often excludes any form of "universal mental dimension." Spinoza, on the other hand, attributes equal dignity to thought and extension, considering them co-original and inseparable aspects of substance. For this reason, his system is closer to a form of rational panpsychism than to pure materialism.

The most profound difference perhaps lies in the ethical and spiritual vision. For Spinoza, knowing the single substance means freeing oneself from sad passions and attuning oneself to the necessary order of Nature. There is no God who rewards or punishes, but a universal ' ' that we can understand and to which we can 'adapt' through reason. Other monisms, while seeking unity, do not always imply this ethical dimension based on understanding and acceptance of the necessary.

The Spinozian concept of Deus sive Natura in the fields of physics and contemporary cosmology.

The Spinozian concept of Deus sive Natura is reflected in various ways in contemporary scientific views.

The universe as a single physical system.

Spinoza saw Nature as a coherent whole, where each part is an expression of the same reality. In modern cosmology, the universe is often described as a unified system governed by universal laws. Einstein's theory of general relativity, for example, proposes a continuous space-time in which matter and energy are manifestations of a single structure. Einstein himself stated that he believed "in Spinoza's God," that is, in a universal order devoid of miraculous interventions but filled with rational harmony.

Interconnection and non-separability.

Quantum mechanics shows phenomena, such as entanglement, in which two particles remain correlated regardless of distance. This recalls Spinoza's view that every entity is part of an infinite network of relationships. In Spinoza's language, one could say that particles share attributes of the same substance and behave in ways that reveal an underlying unity.

Unified fields and fundamental force theory.

Research in theoretical physics, such as unified field theories, seeks to show that the four fundamental forces of nature (gravity, electromagnetism, strong and weak nuclear interactions, and the) are manifestations of a single principle. This approach is very similar to the logic of Deus sive Natura: a single "fabric" that manifests itself in different forms but comes from a single origin.

Quantum cosmology and the concept of totality.
Models such as the cosmology of the "self-consistent" or cyclical universe eliminate the distinction between "beginning" and external "creation." These approaches, in a non-theological way, assume that the universe is "self-caused," autonomous and without the need for an external agent, a concept practically identical to that of Spinozian substance.

Approaches to consciousness as a diffuse property.
Some emerging theories—for example, those connecting consciousness to quantum processes (Penrose, Hameroff)—suggest that the mind is not separate from the physical structure of the universe but deeply rooted in it. This parallel with Spinoza is evident: mind and matter are two sides of the same reality, two attributes of a single substance.

These examples do not imply that physicists directly adopt Spinozian philosophy, but they show how many modern lines of research converge towards a conception of reality as an indivisible whole, governed by necessary laws and not by external interventions — precisely the core of Deus sive Natura.
The following table presents the convergence between Spinozian philosophy and modern science in chronological order.

Table7- Chronological table of convergences between Spinozian philosophy and modern science.

Year / Perio d	Philosophical event (Spinoza)	Scientific event/disc overy	Reference citation	Conceptual connection
1632	Birth of Spinoza.	Galileo consolidate s	*Galileo*: "Nature... does nothing	Premise for the idea of immutable

Year / Period	Philosophical event (Spinoza)	Scientific event/discovery	Reference citation	Conceptual connection
		experimental physics and the principle of uniformity of natural laws.	in vain, and everything is done in order and according to certain laws."	universal laws, basis of *Deus sive Natura*.
1660–1665	Formulation of *Deus sive Natura*: a single infinite substance, cause of itself.	Kepler and Newton develop the mathematical description of motion.	*Spinoza*: "By God I mean an absolutely infinite being, that is, a substance consisting of infinite attributes..." (*Ethics*, I, Def. VI).	Infinite substance and universal laws overlap in the rational view of nature.
1677	Posthumous publication of *Ethica*.	Newton publishes *Principia Mathematica* (1687).	*Newton*: "Hypotheses non fingo" — laws are observed, not invented.	Both see reality as governed by necessary and intelligible causes.
Late 19th century	—	Maxwell unifies electricity and magnetism.	*Maxwell*: "Matter and energy coexist in a unified field."	Analogous to the plurality of attributes of Spinoza's single substance.
1905–1915	—	Einstein formulates special and general relativity.	*Einstein*: "I believe in Spinoza's God, who reveals himself in th I harmony of all that exists."	Space-time as a unified manifestation of the natural order.
1930s – 1960s	—	Quantum mechanics,	*Bohr*: "Everything we call real	Unity and deep interconnectio

Year / Perio d	Philosophical event (Spinoza)	Scientific event/disc overy	Reference citation	Conceptual connection
		entangleme nt.	is made up of things that we cannot consider real in the classical sense."	n, consistent with Spinoza's idea of a necessary network of relationships.
1970s – 2000s	Renewed academic interest in Spinoza.	Theories of grand unification and cyclic cosmology.	*Hawking*: "The very concept of God can be identified with the laws of the universe."	Direct parallel with *Deus sive Natura* as intrinsic law.
Today (2020 – 2025)	Philosophy of mind influenced by Spinoza.	Panpsychist studies and quantum approaches to consciousn ess.	*Spinoza*: "The order and connection of ideas are the same as the order and connection of things."	Integrated view of mind and matter that anticipates theories of consciousness as a universal property.

The following table presents a detailed comparison between Spinoza's monism and Hegel's monism.

Table8. Detailed comparison between Spinoza's monism and Hegel's monism.

Aspect	Spinoza	Hegel
Ontological foundation	Unique, eternal, necessary substance (*Deus sive Natura*).	Absolute Spirit, dynamic and self-developing reality.
Nature of reality	Static, necessary order, without contradictions.	Dialectic: contradiction and synthesis as the engine of evolution.
Mind–matter	Two parallel attributes of the same substance.	Moments in a process leading to absolute consciousness.

Aspect	Spinoza	Hegel
Temporalit y	Timelessness, eternity as the true dimension of being.	Historical, truth that emerges over time.
Ethics	Freedom = understanding of the necessary natural order.	Freedom = historical realization in the institutions of the Spirit.

The influence of Spinoza's formulation of monism on the historical context of the 17th century.

Spinoza's monism was formed in an extremely complex historical context, that of the 17th century, marked by religious and political tensions that had a direct impact on his thinking.

The 17th century in Europe was characterized by the legacy of the Thirty Years' War (1618–1648), a devastating conflict in which religious motivations (between Catholics and Protestants) were intertwined with political ones.

This war highlighted how deep internal divisions based on religious dogma could be and led to an urgent need to base coexistence on universal principles, beyond confessional divisions.

Spinoza responded by proposing a single rational order (Deus sive Natura), where all differences dissolve into a common infinite substance.

In the Netherlands, the Republic of the United Provinces was relatively more tolerant than the more rigid Catholic or Protestant states, but it was not immune to theological conflicts. Amsterdam was a center of religious pluralism: Calvinists, Catholics, Jews, and Mennonites coexisted, but with rigid power hierarchies and strong pressure on internal orthodoxies. In this context, Spinoza, a Sephardic Jew, was excommunicated from his community for his ideas considered heretical, especially his rejection of a personal and transcendent God.

Descartes' rationalist thinking was dominant, with its separation between *res cogitans* (mind) and *res extensa* (matter).

Spinoza initially studied and embraced Cartesian analytical tools, but radically rejected dualism, replacing it with a monistic view in which mind and body are two attributes of the same substance. This choice also responded to a historical need: to provide common foundations for knowledge beyond religious and philosophical divisions.

Moreover, 17th-century Europe was experiencing the expansion of modern science (Galileo, Kepler, Newton). Many philosophers sought to reconcile the new science with traditional religion; Spinoza broke with this pattern and made rationality itself the sole criterion of truth. His monism was a response to cultural and spiritual fragmentation: if everything is a manifestation of the same substance, there is no room for an ontological hierarchy between 'divine' and 'natural' spheres.

In an era of conflicts arising from different interpretations of divine will, Spinoza rejected the idea of an interventionist God and identified God with necessary nature. This move had radical political and religious implications: it eliminated the idea of a God who favors one people or intervenes miraculously, thus reducing the basis for sectarian conflicts.

In short, the historical context of the 17th century, divided by war, confessional tensions, and the emergence of r scientific rationalism, prompted Spinoza to formulate a monism with precise characteristics:

- universal, to overcome religious divisions;
- rational, to base knowledge on necessity and immutable laws;
- inclusive, because mind and matter are manifestations of the same principle.

His Deus sive Natura is, in essence, also a political and ethical program for an era marked by fragmentation.

Spinoza's monism (*Deus sive Natura*) certainly belongs to the family of pantheistic conceptions, but it profoundly modifies some of its fundamental assumptions.

The point of contact with traditional pantheism can be summarized in the concept that, as in pantheism, Spinoza affirms that God is immanent in the universe: there is no transcendent 'beyond', everything that exists is within divine reality. There is no dichotomy between the material world and the spiritual world: everything is part of the same totality.

However, there are structural differences that can be summarized as follows:

Absence of personalization of God.

In traditional pantheism, especially in mystical or religious currents, God is often conceived as a universal consciousness or will that permeates the world. Spinoza rejects the idea of a personal God, endowed with intentions, emotions, or miraculous interventions. For him, God is the necessary order of Nature, eternal and impersonal.

Radical substantial unity.

Many forms of pantheism admit a plurality of divine "manifestations" (spirits, souls, cosmic forces) as distinct parts of being. Spinoza, on the other hand, asserts that everything is a single "substantial being" with infinite attributes; what we perceive as diversity are only "modes" of this substance.

Rational rigor.

Traditional pantheism may have mythical, poetic, or contemplative foundations. Spinoza constructs his system as a geometric demonstration: every thesis on the divine derives from definitions and propositions, not from mystical experiences.

There are also differences between Spinoza's view of pantheism and the classical view in terms of philosophical implications

First of all, there are no exceptions: in Spinoza, the laws of nature are absolute; there is no "divine favor" or "miracle" that suspends necessity.

As for the ethics of knowledge: liberation and bliss derive from understanding the necessary order of the universe, not from worship or devotion.

The following table represents the transition from classical pantheism to Spinozian monism, showing the points of contact and the breaks.

Table9- The transition from classical pantheism to Spinozian monism.

Conceptual element	**Traditional pantheism**	**Point of contact**	**Spinoza's monism**
Vision of God	Immanent but often personal and volitional	God present in the world, not transcendent	Immanent, impersonal, identified with *Nature*
Relationship with the world	God "inhabits" the world, can interact with and modify it	Union between divine reality and natural reality	The world is a "mode" of God/Nature, free from external intervention
Unity vs. multiplicity	Many divine manifestations , spiritual hierarchies	All part of the same cosmic whole	A single substance with infinite attributes, without hierarchies
Laws and miracles	Possible exceptions through miracles or interventions	Recognition of the divine dimension in nature	No miracles: universal and necessary natural laws
Philosophical foundation	Often mystical or religious	Intent to unify the sacred and nature	Rational and geometric method, logical deduction

Conceptual element	Traditional pantheism	Point of contact	Spinoza's monism
Ethical purpose	Harmony with the divine through devotion or contemplation	Seeking unity with the whole	Rational understanding of necessary order as a path to freedom

Does Einstein's interpretation of Spinoza's monism and its implications reflect in his scientific work?

Einstein's interpretation of Spinoza's monism is reflected in the conceptual structure of his scientific work in several key ways. Here are a few:

God as the order of Nature.

Einstein rejected the idea of a personal, interventionist God, claiming to believe in "Spinoza's God," that is, in the necessary harmony that governs the universe. For him, divinity coincides with universal laws, unchanging and independent of human will, a concept perfectly in line with *Deus sive Natura.*

In his studies on relativity, Einstein showed how space and time are aspects of the same entity (space-time) governed by geometric laws. There is no room for arbitrary interventions, because everything follows mathematical necessity.

The unity and coherence of the whole.

Spinoza argued that there is a single substance with infinite attributes; the separation between *res cogitans* and *res extensa* is illusory.

Einstein, similarly, always sought to unify seemingly distinct phenomena under common laws (for example, light and electromagnetism, time and space, matter and energy).

The famous equation E=mc^2 expresses precisely this substantial unity between mass and energy, similar to Spinoza's idea that diversity is only a different "modality" of the same principle.

The necessity of natural laws.

For Spinoza, events happen according to necessity; 'miracles' are impossible because they would violate the order of divine substance. Einstein saw physical laws as absolute and universal, valid everywhere, without exception. General relativity does not allow for 'anomalies' caused by divine will or supernatural events: the curvature of space-time obeys precise mathematical equations.

The ethics of knowledge.

Spinoza linked freedom to a rational understanding of the universal order. Einstein shared the idea that the study of nature was a path to understanding and harmony with the cosmos, avoiding superstition and dogmatism.

The following table chronologically traces the transition from Spinoza's philosophical formulations to their resonance in Einstein's physical models according to this synthetic conceptual flow:

- Spinoza defines *Deus sive Natura* (necessary order).
- The Enlightenment consolidates the rational and anti-supernatural approach.
- Classical physics begins the search for universal and unifying laws.
- Einstein makes a leap in the understanding of physical unity (relativity).
- Modern cosmology extends the idea of universal order to a cosmic scale.

Table10- Connection between Spinoza's philosophical formulations and scientific evolution.

Period	Stage	Philosophical/scientific content	Conceptual connection
17th century (1660 – 1677)	Spinoza – Monism (*Deus sive Natura*)	A single infinite substance, God and Nature are the same thing; universal and necessary laws; rejection of a personal God.	Foundation: the order of Nature is divinity itself; concept of unity and necessity.
Late 17th – 18th century	Enlightenment and scientific rationalism	Centrality of reason, rejection of the supernatural, natural laws as an explanation of the universe.	Amplification of Spinoza's rational approach in the scientific field.
19th century	Classical physics and electromagnetic unification (Maxwell)	Mathematical description of electromagnetism; first unifications of physical phenomena.	Example of the "unity of manifestations," consistent with Spinoza's principle of a single substance.
1905 – Special relativity	Einstein	Unification of space and time in space-time; equivalence between mass and energy (E=mc2).	Direct reference to substantial unity: apparent diversity as different "modes" of the same reality.
1915 – General relativity	Einstein	Gravity as curvature of space-time; invariable universal laws.	Reflection of Spinoza's concept of necessary order, without exceptions or miracles.
Mid-20th century	Physical cosmology	Big Bang model, expansion of the universe, universal cosmological laws.	Confirmation of immanence: everything is governed by natural laws that correspond to the "harmony" of God-Nature.
21st century –	Particle physics and	Search for a Theory of Everything (quantum gravity,	Scientific continuation of Spinoza's idea of a single

Period	Stage	Philosophical/scientific content	Conceptual connection
Modern cosmology	unification theory	unification of fundamental forces).	substance with infinite attributes.

Mind and body as attributes. The parallel vision.

When Baruch Spinoza, in a modest house in The Hague in 1675, was putting the finishing touches to his *Ethica*, he was well aware that he was challenging Descartes' dualism. Descartes had clearly distinguished *res cogitans* (thinking substance) from *res extensa* (corporeal substance). For Spinoza, that division was artificial. He argued that mind and body are not two separate substances, but two attributes of the same unique reality: Substance, which he called "God or Nature."

This is the heart of his parallelism. What happens in the body has a correspondence in the mind, and vice versa. There is no direct causality between the two, because they are not instruments that influence each other from the outside: they are two sides of the same order. Spinoza writes, with the sobriety of his geometric axioms: "*The order and connection of ideas is the same as the order and connection of things*."

Let us imagine the philosopher in the 1660s, reading and translating scientific and theological texts in several languages. He embraces Euclidean geometry and applies it to thought. Spinoza lives in a community of religious exiles and frequents mathematicians, doctors, and opticians. Working with a microscope, he polishes lenses for scientific instruments: a precision activity that, , reflects his intention to observe reality without separating mind and matter.

This is not pure abstraction. Spinoza reduces the distance between philosophy and life. Accounts from the time speak of his calmness in heated discussions with Calvinist theologians. Spinoza argues that the human mind is a mode of the attribute of thought, just as the body is a mode of the attribute of extension. Both are modes of the same substance, subject to the same necessary laws.

This view had a decisive influence on modern thought. In the 20th century, philosopher and psychologist William James, reflecting on the mind-body relationship, recognized Spinoza as a precursor to the idea that consciousness and the physical world are aspects of a single reality. Years later, neuroscientist Antonio Damasio would quote Spinoza when discussing *the "errors of* dualism" and the need for a science of the mind rooted in biology.

Spinozian parallelism also resonates in today's language of neurocognition. Studies of neural networks show patterns in which the "flow of ideas" mirrors the "flow of bodily signals," perfectly echoing the "*ordo et connexio rerum et idearum*." The mind does not float in an isolated ether: it emerges from the body and forms a single system with it.

In his time, this conception dismantled the theological motif of the body as a prison of the soul. For Spinoza, the mind is not condemned to govern a resistant body, nor is the body an obstacle to the intellect. They are one life, observed with two different conceptual instruments. It is the same phenomenon seen from two perspectives.

When we talk about panpsychism today, Spinoza's thesis appears as a classic example of how consciousness and matter can be inseparable. And, in his dry language, Spinoza gives us an idea that is still relevant today: to understand the body is to understand the mind, and to understand the mind is to understand the body. Neither path is secondary to the other.

Spinoza's view and other theories on the mind-body relationship.

Spinoza's parallelism is clearly distinct from both Cartesian dualism and materialism.

For Spinoza, mind and body are not two separate entities that interact, but two attributes of the same unique substance—Deus sive Natura.

This means that every mental event has a corresponding physical event, and every physical event has a corresponding mental event. This correspondence does not imply direct causality between mind and body, but rather a structural synchrony: "*the order and connection of ideas is the same as the order and connection of things*."

Descartes distinguishes between mind (res cogitans) and body (res extensa), two autonomous substances. In the Cartesian model, the mind can cause movements of the body and the body can cause sensations in the mind, through a point of interaction (for Descartes, the pineal gland).

Spinoza rejects this causal interaction: mind and body do not influence each other because they are two ways of describing the same reality. In practice, Spinoza avoids the problem of the "mechanisms" of connection between different entities that has plagued dualism.

As for materialism, it is well known that it reduces everything to physical phenomena. What we call mind is, in this view, an epiphenomenon or the product of neurobiological processes. Spinoza, on the other hand, attributes to thought an ontological status equal to extension. He does not reduce consciousness to matter, nor matter to consciousness: both are manifestations of a single substance.

This position is close to modern forms of panpsychism: consciousness would be a fundamental and widespread aspect of reality, not an 'emergent function' exclusive to complex brains.

Spinoza's parallelism finds echoes today in neuroscience, which proposes models that describe the mind as a projection of bodily processes without ontological separation.

Similarly, it is echoed in theoretical physics and the philosophy of mind, based on monistic or panpsychist hypotheses that attempt to overcome reductionism and dualism.

Below, I have attached a table that summarizes the conceptual and operational differences between Spinozian parallelism, Cartesian dualism, and materialism.

Table11- Conceptual and operational differences between Spinozian parallelism, Cartesian dualism, and materialism.

Aspect	**Spinozian parallelism**	**Cartesian dualism**	**Materialism**	**Historical examples and figures**
Mind-body nature	Two attributes of the same substance (*Deus sive Natura*).	Two distinct substances: mind (*res cogitans*) and body (*res extensa*).	The mind is a product of the physical processes of the body.	Spinoza, Leibniz (pre-established harmony), Goethe.
Causation	No direct causality: structural correspondence and synchrony.	Direct interaction between mind and body (e.g., pineal gland).	Physical causality only: the mind as an epiphenomenon of bodily processes.	Descartes, Malebranche (occasionalism), Gilbert Ryle (criticism of the "ghost in the machine").
Ontological status	Mind and body have equal ontological dignity.	The mind is ontologically superior to the body.	The body is primary; the mind is secondary or emergent.	Spinoza, Schelling (monism), Democritus (atomism).
Philosophical implication	Monistic view: unity of existence and substance without separation.	Need for a "bridge" between two different substances.	Reductionist view: I denial of an autonomous mental plane.	Epicurus (ancient materialism), Marx (historical materialism), Daniel Dennett

Aspect	Spinozian parallelism	Cartesian dualism	Materialism	Historical examples and figures
				(contemporary materialism).

A difficult overlap.

For Spinoza, consciousness is not a privilege reserved for humans. It is a constitutive attribute of substance, present in all things in proportion to their complexity. A stone, a plant, a cat, and a human being participate in the same universal life. The difference is not one of nature, but of degree. This view anticipates the perspective we now call panpsychism: the idea that consciousness, or at least some form of subjective experience, permeates everything that exists.

This premise can be said to be accurate, but a small distinction must be made to avoid completely superimposing Spinoza's thought on modern panpsychism.

Spinoza considers thought and extension as two parallel and necessary attributes of the same infinite substance (Deus sive Natura). This means that every entity, by the mere fact of 'existing' as a mode of substance, has a mental correspondence or 'idea' — therefore, in a sense, a form of consciousness or representation. However, this consciousness is not autonomous, individual in the modern sense, but part of a single mental order, parallel to the physical order of the universe.

Contemporary panpsychism, on the other hand, tends to say that everything, even the simplest particle, possesses a form of consciousness or experience of its own, albeit extremely rudimentary. Here, consciousness is distributed as an intrinsic property of each separate entity.

In Spinozian pantheism, on the other hand, there are no separate entities with isolated consciousnesses: everything participates in a single substance, and the order of ideas mirrors the order of things. The difference between a stone, a plant, and a human being is not qualitative in the nature of consciousness, but quantitative in the complexity and relationships of the ideas that constitute them.

Therefore, the greatest similarity with panpsychism is the idea that 'everything thinks' or 'everything has a mental aspect'.

The substantial difference, however, lies in the fact that Spinoza sees this thought as part of a unified order linked to a single divine substance, not as a mosaic of independent consciousnesses.

Consciousness as a universal property.

Spinoza writes in Ethics that "*the order and connection of ideas is the same as the order and connection of things*." There is no ontological hierarchy between thought and extension. Thinking and, in a broad sense, feeling, are part of the same network of necessity. In this scheme, human beings do not dominate Nature from the outside, but belong to it like a cell to an organism.

The ethical implications are profound. If every entity participates to some extent in the light of consciousness, then respect for living beings and the environment is not just a moral choice or a social convention, but a logical consequence of the structure of the universe. In the 20th century, figures such as Albert Schweitzer took up this thread, formulating "respect for life" as a universal principle. Mohandas Gandhi, inspired by ahimsa, also put this insight into practice: non-violence is not only about human action, but a global attitude towards all forms of existence.

There is an anecdote that illustrates Spinoza's spirit well: it is said that, walking among the dunes near The Hague, he loved to observe everything as part of the same divine necessity, from the flight of birds to the ripples of the sea. For him, contemplating a grain of sand had the same value, in terms of truth, as contemplating the starry skies. This ontological equality already erased anthropocentric presumption at that time.

Today, neuroscience and theoretical physics are once again questioning consciousness as a widespread property. Some models, such as Giulio Tononi's Integrated Information Theory or Roger Penrose's theories on quantum consciousness, propose a continuous field of experience that extends beyond the human brain. These are not demonstrations of Spinoza's theses, but unexpected echoes of an intuition formulated more than three centuries ago.

Spinoza's vision requires us to think of ethical responsibility as coextensive with being. We do not act on an inert world: we act in a living world, of which we are an expression. From this perspective, destroying a forest or poisoning a river is not just ecological damage. It is, to use Spinoza's language, a diminution of the very power to exist of the substance of which we are a part.

The vision of a universal consciousness and the presence of evil in the world, according to Spinozian ethics.

According to Spinoza, the question of evil and suffering does not arise in the same way as it is understood in common theological or moral tradition. In his ethics, the "universal consciousness" (attributed to the single substance that is Deus sive Natura) is not a "moral eye" that judges according to absolute categories of good and evil.

For Spinoza, good and evil are not intrinsic properties of the world, but relative ways in which human beings evaluat l things in relation to their capacity to exist and act, that is, their "*potentia.*" "Evil" is anything that diminishes "potentia"; "good" is anything that increases it. From this point of view, suffering and destruction are not the whims of a God-person nor anomalies in a perfect order, but necessary effects of the infinite causes and interactions that make up reality.

Universal consciousness, being coextensive with substance, is not selective: it also includes those phenomena that we call pain, illness, and death. But it sees them as part of the same fabric of connections that produces life, growth, and joy. Spinoza often uses the image of the natural order: a storm can knock down a tree, but it is not "evil"; it is only the consequence of the laws of Nature.

Ethically, this view shifts the focus from judgment to understanding. If human beings want to act according to reason, they must interpret evil and suffering as necessary data and seek to transform the conditions that generate them into forms of greater harmony and common power. Albert Schweitzer, totally different in terms of time and education, expressed a similar perspective with "respect for life": it is not a question of eliminating all pain (which is impossible) but of reducing what is avoidable, increasing the possibilities of existence and freedom for all.

In Spinozian terms, when a harmful action destroys part of the substance (a forest, a community, an individual), it is not 'punished' by a separate God, but reduces the overall power of the system of which we are a part. Ethics therefore arises from the awareness that acting against this balance means acting against ourselves.

'NATURAL EVIL' AND UNIVERSAL CONSCIOUSNESS.

Spinoza's idea of a universal consciousness, an integral part of the single substance Deus sive Natura, radically transforms the way we can interpret 'natural evil' (earthquakes, diseases, predation, psychological suffering) not as moral anomalies, but as events internal to the necessary structure of reality.

In the philosophy of mind, this perspective dialogues with positions that today seek to free the concept of consciousness from its exclusive link to the human brain. Some theorists, such as Galen Strawson, argue that if consciousness is a fundamental aspect of matter, then what we perceive as "evil" or "pain" is also part of an experiential continuum, coextensive with nature.

From a neuroscientific point of view, models such as Giulio Tononi's "*Integrated Information* Theory" propose that consciousness is an emergent property of a system's ability to integrate information. In this key, destructive events (from neuronal degeneration to the disruption of ecological networks) locally reduce the "quantity" of consciousness distributable in that space-time. These are not 'punishments', but decreases in the structural complexity that makes experience possible.

The discussion on "natural evil" is enriched when so-called quantum consciousness comes into play, with hypotheses such as those of Roger Penrose and Stuart Hameroff, according to which the quantum processes underlying matter are carriers of an elementary potential for experience. In this scenario, suffering is not the absence of consciousness, but a particular configuration of it, determined by constraints and interactions that are inevitable on the scale of the universe.

On an ethical level, the convergence between Spinoza and these theories leads to an uncomfortable but powerful conclusion: if everything participates in universal consciousness, 'natural evil' demands a response not of metaphysical condemnation, but of practical transformation. For the contemporary philosopher or researcher, this means studying how to reduce the states that diminish the power to exist, whether it is preventing a human epidemic, restoring a destroyed habitat, or intervening in a social network that produces alienation.

In the cultural sphere, this line of thinking has influenced movements such as ecocentrism and evolutionary bioethics, which see the increase in the "overall power" of the ecosystem and human societies as the true antidote to suffering. This is a direct echo of Spinoza, who taught that understanding the causes is the first step towards increasing the positive effects.

The following table presents an immediate and coherent picture of the correspondences between Spinoza's view of natural evil and the main contemporary neuroscientific theories:

Table12- Correspondences between Spinoza's view of natural evil and the main contemporary neuroscientific theories:

Spinoza's explanation of natural evil	**Contemporary neuroscientific theories**
Evil is not an autonomous entity, but the necessary effect of infinite causes acting according to the laws of Nature.	Phenomena such as earthquakes, epidemics, or neuronal degeneration are interpreted as inevitable expressions of the complexity and dynamics of life.
Suffering reduces the *potency* of existence; good increases it. Ethics aims to understand in order to transform.	Models such as Integrated Information Theory (Giulio Tononi) measure consciousness in terms of information integration; biological or environmental damage reduces this integration and therefore the quality of experience.
There is no transcendent moral judgement: ethics means increasing the power and harmony of universal substance.	Neuromodels and therapeutic interventions aim to reduce mental and physical states that limit well-being, expanding cognitive and emotional capacities.
Responding to natural evil means acting to increase common power: cooperation, respect for all forms of existence, ecological balance.	Neuroscientific and psychosocial research develops strategies to prevent trauma, disease, and environmental degradation, promoting resilience and collective well-being.

Theories on quantum consciousness and Spinoza's anticipations on the relationship between mind and matter.

Modern theories on quantum consciousness show both points of convergence and divergence with respect to Spinoza's insights on the mind-matter relationship. Here is a summary of some points of convergence:

Consciousness as a universal property.

Spinoza sees consciousness (or rather, the order of ideas) as an attribute of the single substance 'Deus sive Natura', present in all modes of being, even the simplest ones. The hypotheses of Penrose and Hameroff—according to which fundamental quantum processes of matter can generate conscious states—are close to this view, because they postulate that consciousness is rooted in the ultimate structures of reality, not limited to the human brain.

The mind-matter interconnection.

Spinoza's assumption that "the order and connection of ideas is the same as the order and connection of things" implies a structural parallelism between mental and physical events. Quantum theories hypothesize that mental phenomena emerge from deeply correlated physical processes, maintaining in some way a "shared order" — for example, through states of quantum coherence in neuronal microtubules.

A question of degree, not nature.

Spinoza argues that differences in consciousness between living beings are quantitative, not qualitative. Similarly, some versions of quantum consciousness and panpsychism see each physical system as carrying a different degree of "experience" or "proto-consciousness."

These points are counterbalanced by some points of divergence:

Unique ontology vs. physicalist ontology.

Spinoza starts from a rigorous monism: mind and body are two modes of the same order, two attributes of substance.

Quantum theories often maintain a methodological distinction between matter and consciousness, seeking to derive the latter from the former, rather than seeing them as two inseparable sides of a single reality.

Necessity vs. probability.

For Spinoza, every event—mental or physical—occurs with necessity according to the laws of Nature. Quantum physics, on the other hand, introduces an element of probabilistic indeterminacy, which in its formalism seems incompatible with Spinoza's absolute determination.

Absence of ethics.

The theories of Penrose, Hameroff, and others do not necessarily imply a universal ethical dimension. In Spinoza, on the other hand, the recognition of consciousness as a universal attribute has precise moral consequences: respect for all forms of existence and the increase of common "power."

In summary: quantum consciousness, in its most radical formulations, takes up Spinoza's suggestion of a mind diffused throughout reality, but moves on empirical and physicalist bases. If Spinoza already 'sees' a universal mind through philosophy, contemporary science seeks to verify it through models and experiments.

The role of Eastern philosophy, particularly Buddhism and Hinduism, in the historical development of panpsychism.

We can consider the contribution of Buddhism and Hinduism to panpsychism as a cultural and philosophical thread that spans millennia, and thus we can compare it with the rigorous structure of Spinozian monism.

In Buddhism, three fundamental aspects can be highlighted. The first is universal interdependence (*Pratītyasamutpāda*). In

Buddhism, every phenomenon exists in relation to all others. This interdependent network also extends to consciousness, which is not an isolated property of the individual but a distributed process, born from the interaction of infinite causes and conditions.

The second aspect is non-dual consciousness. In some Mahāyāna schools, such as Zen and Yogācāra, the mind is seen as fundamentally non-separate from phenomenal reality: 'consciousness' permeates all forms, albeit in ways that do not imply a permanent self.

The third important aspect is the ethic of universal compassion. The belief that all sentient beings participate in this dimension of consciousness leads to an ethical imperative similar to that of Spinoza: to reduce suffering and ignorance, promoting harmony as a whole.

In the Vedantic philosophy of Hinduism, we find the concept of Ātman–Brahman. Every individual consciousness (ātman) is a manifestation of the universal absolute principle (Brahman). This fundamental identity is very close to Spinozian monism, where every mind is a "mode" of the single substance.

In the Purāṇas and in many Śivaite and Vaiṣṇavite schools, even natural elements such as rivers or mountains are bearers of life and consciousness, not necessarily anthropomorphic, but part of a sacred continuum.

There is also an analogy in the hierarchy of degrees of consciousness. Although consciousness is universal, it is often conceived as manifestations of varying intensity or purity, a concept that echoes Spinozian "gradualism."

The following table summarizes the historical evolution of panpsychism in chronological order, starting from its roots in Hinduism and Buddhism to Spinoza and contemporary theories of consciousness.

Table13- Chronological summary of the historical evolution of panpsychism, starting from its roots in Hinduism and Buddhism to Spinoza and contemporary theories of consciousness:

Era / Tradition	Main figure or school	Vision of consciousness	Influence on later panpsychism
~1500 BC – Vedic and Vedantic Hinduism	Vedic and Upanishadic texts	Individual consciousness (*ātman*) identical to universal reality (*Brahman*); the entire universe is a manifestation of an eternal principle.	Provides a sacred monistic structure, in which everything is part of a continuum of consciousness.
~500 BCE – Early Buddhism	Buddha and Theravāda schools	Consciousness as an interdependent process (*Pratītyasamutpāda*), devoid of a permanent self; universal compassion towards all sentient beings.	Introduces the idea of diffuse and impermanent consciousness, with ethical-universalistic implications.
~200 CE – Mahāyāna Buddhism	Yogācāra, Madhyamaka	Non-dual universal mind or *śūnyatā*; all phenomenal forms participate in the dimension of consciousness.	Refines the concept of universal consciousness in a non-substantialist key.
1600–1677 – Spinoza	European rationalist philosophy	Strict monism (*Deus sive Natura*); mind and body as two inseparable attributes of substance; consciousness present in every mode of being.	Integrates the universal principle into a deterministic and logical-mathematical framework, with strong ethical implications of harmony and common empowerment.
1900–1970 – Modern philosoph	James, Russell, Whitehead	Consciousness as a fundamental principle of reality; every entity	Reformulates insights similar to Eastern and Spinozian

Era / Tradition	Main figure or school	Vision of consciousness	Influence on later panpsychism
ical panpsychism		has an elementary form of experience.	traditions in a scientific and analytical I key.
1970–present – Contemporary theories of consciousness	Integrated Information Theory (Tononi), Quantum consciousness (Penrose/ Hameroff)	Models that attribute degrees of consciousness to complex physical systems; quantum hypotheses that see consciousness as rooted in the fundamental processes of matter.	Attempts to empirically verify the concept of universal consciousness and quantify it.

Deus sive Natura as a scientific paradigm.

Spinoza wrote Deus sive Natura in the 17th century, in a Europe swept by the scientific revolution. Galileo had died a few decades earlier. Newton was formulating his *Philosophiae Naturalis Principia Mathematica*. But Spinoza, shut away in his room in Amsterdam, polished lenses and thought. He thought big. He imagined a universe where God is not a separate being who intervenes from above, but the very substance of the cosmos, infinite and necessary.

In this vision, physical reality and mental reality are two faces of the same entity. Spinoza calls these two faces "attributes" of substance: extension and thought. What exists is always and simultaneously *res extensa* and *res cogitans*. There is no logical void between matter and mind. This is already, in some ways, a preview of the unified vision pursued by modern theoretical physics.

Today, physicists are searching for a "theory of everything" that unifies the fundamental forces and reconciles quantum mechanics with general relativity. Spinoza, without mathematical formulas, had already conceived of a universe without ontological gaps. Every phenomenon follows immutable rules, whose necessity is written into the very structure of substance.

Many scientists of the twentieth century perceived this harmony. For Einstein, Spinoza's Deus sive Natura corresponded to the beauty of physical laws: not a watchful and moral eye, but a necessary intertwining of causes and effects, similar to the curvature of space-time.

If Spinoza had been able to observe the theory of relativity, he would probably have recognized his idea in the geometric structure of the universe: substance manifesting itself as space, time, and energy, inseparable and governed by rigorous laws. Similarly, in quantum mechanics—where matter and energy appear as manifestations of fields and probabilities—we find an echo of his monism: not two realities, but one, which appears in different forms depending on the point of view.

In the modern laboratory, this unit has concrete faces. In the silent halls of CERN in Geneva, particles are born and disappear in infinitesimal fractions of time. For the theoretical physicist, they are manifestations of a universal "field." For Spinoza, they would be modes of substance, transitory but necessary. In both cases, there is no separation between "creation" and "functioning": becoming is part of being itself.

Even the most current theories on quantum consciousness unknowingly echo Spinozian nuances. If consciousness is seen as emerging from deep quantum processes, it is not something external to matter, but a form of its organization. Spinoza had already intuited this: the mind is an attribute of substance, intertwined with extension. The language changes, but the concept remains close.

The 'Deus sive Natura' paradigm is therefore more than philosophy. It is a framework in which modern physics can feel at home. Its relevance today can be seen in the drive towards a single coherent description of the universe. There is a direct link between Spinozian pantheism and the 'unification of forces' discussed by theoretical physicists: the belief that behind the variety of forms there is a single fabric.

And perhaps, if a researcher today cycles through the canals of Amsterdam and passes the house where Spinoza wrote, he can imagine that this thinker, isolated among books and optical instruments, was laying the philosophical foundations for one of the most radical scientific ideas of our time: the universe as a single, infinite substance of which we are a part.

Spinozian monism and the search for the theory of everything.

When Spinoza imagines the single substance, Deus sive Natura, he does not think in terms of separate compartments. There is no realm of matter opposed to that of the mind, nor a physical world separate from a spiritual one. There is only one principle, infinite and necessary, which manifests itself in different forms. It is a philosophical intuition of the 17th century but, surprisingly, it corresponds to a specific goal of contemporary theoretical physics: to describe the universe with a single language, a 'theory of everything' capable of uniting the laws that govern the cosmos, from the subatomic world to the expansion of galaxies.

Over the last fifty years, this research has developed in different directions. One of the most decisive was the formulation of the Standard Model of particles (1960s-1970s), which brings together three of the four fundamental forces—electromagnetism, weak force, strong force—into a coherent framework, leaving gravity out for now. In a Spinozian sense, this model is a partial representation of the singl l substance: many manifestations united under the same scheme, but not yet complete.

The challenge of gravity has become the field in which monistic ideas resonate most strongly. Einstein's theory of general relativity, which describes space-time as an elastic fabric deformed by matter and energy, is conceptually consistent with Spinoza's vision: "substance" is not a static container, but the very structure of being.

In the 1980s, physicist Brian Greene and other theorists advanced string theories, which imagine all particles and forces as vibrations of a single fundamental entity. Here too, the similarity is evident: different manifestations perceived as separate, but in reality modes of the same reality.

Another, more recent frontier is that of loop quantum gravity and emergentist approaches, according to which space and time are not fundamental, but emerge from deeper quantum relationships. In Spinozian terms, we could say that extension and thought—or their modern equivalents—are different ways of grasping a universal fabric that lies "behind" apparent forms.

Also in recent decades, interest in consciousness as an integral part of physical reality has opened up a surprising avenue. Theories such as *Integrated Information Theory* (Tononi) or quantum consciousness hypotheses (Penrose and Hameroff) see the mind not as a separate property, but as a component inscribed in natural laws. It is an almost literal return to Spinoza's vision: mind and matter as inseparable attributes of a single substance.

Today, in the corridors of CERN, in gravitational wave observatories such as LIGO, or in quantum simulation projects, people are working with the same tension that animated the Amsterdam thinker: to find a unified description of reality, capable of including all phenomena. Spinoza's words — "The order and connection of ideas is the same as the order and connection of things" — could easily be the motto of a group of theoretical physicists seeking to merge equations of quantum mechanics with those of relativity.

The difference lies in the language and tools. Spinoza had logic and geometry. We have particle accelerators and supercomputers. But the core of the endeavor is identical: to reveal the single face of Nature, which is at once matter and mind, particle and field, energy and consciousness.

VII. From the Enlightenment to modern philosophy .

The transition from the 17th to the 18th century marks a turning point for European thought. Spinoza's motto, Deus sive Natura, continues to resonate beneath the rationalist surface of the Enlightenment, but it is transformed by new questions about perception, life, and consciousness. Panpsychism, which in previous eras was intertwined with metaphysics and natural theology, now confronted the precise language of modern science and the desire to order nature under universal laws.

In 1714, mathematician and philosopher Gottfried Wilhelm Leibniz proposed the theory of monads: small indivisible substances, each carrying perception. Monads do not communicate with each other. Each reflects the universe in its own way, like a fragment of a mirror. It is a vision that, while part of the rationalism of the time, restores conscious dignity to matter. The echo of this idea would reach contemporary debates on consciousness as a widespread property.

In late 18th-century Germany, Johann Wolfgang von Goethe moved between poetry and natural science. Goethe did not see nature as an inert mechanism, but as a living organism. In his botanical studies, conducted with almost anatomical precision, he sought to capture the 'original plant' (Urpflanze) that embodies the form of all others. In this intertwining of art and romantic science, consciousness is not exclusive to humans, but a principle diffused throughout the living order.

Friedrich Wilhelm Joseph Schelling takes the argument further. In his Philosophy of Nature, the identity between spirit and nature is total: they are not two separate orders, but the same substance seen from two perspectives. Nature is 'sleeping spirit', and spirit is 'awakened nature'. This idea, which developed in the early 19th century, influenced both the nascent phenomenology and certain mystical currents that ran through Romanticism.

In the 20th century, Edmund Husserl's phenomenology took up the theme again. Husserl studied consciousness as an intentional field, always oriented towards something. Maurice

Merleau-Ponty, in the post-war period, expanded this vision: the body itself becomes the primary locus of consciousness. It is no longer a mind that thinks 'from outside' nature, but a lived experience that takes place in the world. This perspective reopens the dialogue with panpsychism, because consciousness is seen as an embodied phenomenon, diffused in concrete reality.

Finally, Henri Bergson offers a last great fresco at the beginning of the 20th century. His "*élan vital*," or "vital impulse," interprets time as a current of universal consciousness. For Bergson, life is a single creative explosion that runs through matter, governed not only by mechanical cause, but by an internal impulse. Time, far from being a sequence of measurable moments, becomes the living fabric of the universe.

From the Enlightenment to modern philosophy, panpsychism loses its purely metaphysical aura and confronts science, art, and phenomenology. Languages change, but the idea that consciousness is not an isolated island in the human mind, but rather an underground ocean that runs through the totality of reality, remains intact.

Leibniz and conscious monads. Metaphysics of perception.

At the beginning of the 18th century, Europe thought in terms of geometry and astronomy. Isaac Newton described a mechanical universe, made up of bodies that attract and repel each other in space. In Germany, however, Gottfried Wilhelm Leibniz proposed a different model. Born in Leipzig in 1646, Leibniz was a mathematician and diplomat who spent most of his life traveling between courts and libraries. While calculating the binary system and inventing the mechanical calculator, he pondered a more subtle enigma: if matter is made up of increasingly smaller parts, what is the ultimate reality?

In 1714, in his "*Monadology*," Leibniz described the world as a community of indivisible entities: monads. No monad has physical extension. Each monad is a center of perception, a "small living world" that reflects the universe from a unique point of view. It does not interact with others, because everything is already harmonized by God in a "pre-established harmony." Matter, in Leibniz's view, is animated by these invisible and infinite nuclei of consciousness.

Leibniz uses a famous image: a gigantic machine.

This is one of his most famous metaphors to clarify his position on consciousness and to criticize the idea that it can simply emerge from a set of material mechanisms. Here's how the reasoning works:

Leibniz asks us to imagine a machine so large that we can enter it, as if it were a mill or an enormous device. This machine processes information, 'grinds' and 'transforms' — in modern terms, we might think of a brain or a sophisticated physical system.

If you walked among its gears, all you would see would be parts moving other parts, mechanisms pushing, turning, unlocking, or transmitting movement. Even though the machine performs complex functions, you would never find 'consciousness' anywhere, as if it were a locatable component.

Leibniz's point is that mechanical processes, however complex, cannot explain subjective experience: the sensation of 'seeing red', feeling pain, remembering a face, or being self-aware. According to him, these qualities belong to the dimension of 'monads' — fundamental entities endowed with perception — and cannot be reduced to physical movements.

In short, with this metaphor Leibniz tells us that mechanics explains 'how' something works, but not 'what it feels like' from the inside. Consciousness is not just another cog in the wheel, but a completely different dimension that cannot be reduced to pure material movement.

Here, the metaphysics of perception merges with panpsychism: every entity, from a grain of sand to a human being, possesses a degree of perception, even if imperceptible.

This thought dialogues with theology, but anticipates concepts of modern physics. In the monad, we find the idea of a "package of information" that contains an entire state of the universe. Later philosophers, such as Fechner and Whitehead, will use similar images to discuss diffuse consciousness.

Leibniz did not work in isolation. In Paris, he debated with Antoine Arnauld, the heir to Cartesianism, who criticized him for his "too vivid" view of matter. In London, he spoke with scientists from the Royal Society about the idea that natural phenomena are described by mathematical laws but also by metaphysical principles. Eighteenth-century Europe, still steeped in religion, saw his theory as a bridge between faith and reason.

Many interpreters, especially in the Romantic era, read monads as "atoms of spirit." In reality, Leibniz does not identify their perception with human consciousness. A monad can perceive without being "conscious" as we understand the term.

However, each monad contributes to the cosmic symphony. In this way, Leibniz expands the concept of perception beyond the boundaries of the individual mind.

The metaphysics of perception that began with Leibniz remains one of the most refined visions of panpsychism. It transforms the universe from an inert object into a living fabric, in which every point—invisible and indivisible—is already charged with an entire world.

The relationship between Leibniz's 'pre-established harmony' and modern theories of synchronicity.

Leibniz's concept of pre-established harmony and modern theories of synchronicity, especially those formulated by Carl Gustav Jung, arise in very different historical and cultural contexts, but share a basic idea: the universe follows an order that can manifest itself in surprising connections, even when there is no direct causal link between events.

For Leibniz, each monad reflects the entire universe from its own point of view and follows its own "internal program" set by God. Monads do not influence each other, but move in perfect harmony thanks to an original synchronism: pre-established harmony. This means that the order of each being's experiences is already designed to correspond, moment by moment, to that of all others. It is a compact and determined vision, where the coherence of the universe is guaranteed a priori.

Jung's synchronicity, on the other hand, emerged in the 20th century, an era marked by psychoanalysis, quantum physics, and symbolic thinking. Jung describes synchronicity as "meaningful coincidences" between inner facts (such as dreams, intuitions, emotional states) and outer events, without a causal link. What unites the two levels is the perceived meaning. In other words, synchronicity reveals an order that emerges in the observer's consciousness , rather than in a determined structure of the universe.

The bridge between Leibniz and Jung passes through a common idea of global interconnection. For Leibniz, two events that seem to happen "in parallel" do so because their harmony has been defined since the beginning of time. For Jung, they happen because there is an archetypal principle or symbolic field that connects mind and world in a non-causal way. In both cases, the observer perceives an order that transcends the linear mechanics of events.

There are other interesting analogies with certain approaches in contemporary physics. Leibniz's concept of pre-established harmony is reminiscent of a perfectly synchronized system such as a network of atomic clocks: none of them transmits a signal to the others, yet they all show the same time. Jung's synchronicity is closer to the notion of emerging correlations in complex systems, where meaning arises from spontaneously formed relationships.

The cultural thread that unites them is their belief in an order that is 'other' than material causality. In Leibniz, this order is divine will. In Jung, it is a shared archetypal structure, which can also be interpreted in the light of modern hypotheses on the unified field or the extended mind. Both, however, ask the reader to accept that the boundary between inner perception and the external world is not as clear-cut as it appears.

Leibniz's panpsychism and Eastern conceptions of universal consciousness.

Leibniz's panpsychism and Eastern conceptions of universal consciousness share common ground: the belief that reality is permeated by a vital or conscious principle, present everywhere and in everything. However, the roots and conceptual structures of these views are very different.

Leibniz, from the European perspective of the 17th and 18th centuries, imagines a universe made up of infinite monads, indivisible centers of perception. Each monad does not communicate with the others, but acts in perfect correspondence thanks to the pre-established harmony set by God. Consciousness, in this framework, is not something that arises from material combinations, but an intrinsic quality of being, distributed across multiple levels of intensity, from the most rudimentary to the most complex. Here, the organizing principle is personal and transcendent: God sets all monads in motion and coordinates their perceptions.

In Vedic and philosophical India, the concept of Brahman performs a similar function but with different assumptions. Brahman is the absolute, the universal consciousness that exists as the basis of all that appears. The individual, through the inner self (Atman), is not separate from Brahman: the microcosm and the macrocosm coincide, not because of an imposed synchronism, but because they are in reality the same thing. Here there is no 'ordering' God distinct from the world, but an undifferentiated principle that is already the world itself.

In Taoist thought, the Tao is the way and the principle that permeates every phenomenon. It is a spontaneous order, not fixed by an initial act but flowing, changing, without separation between spirit and matter. Unlike Leibnizian monads, each of which retains a unique and immutable point of view, in the Tao every configuration transforms, revealing a dynamic harmony.

In terms of consciousness, Leibniz sees a multiplicity of conscious centers operating in parallel under a fixed design. Vedic India and Taoism, on the other hand, tend to dissolve multiplicity into a single principle: not many isolated "small consciousnesses," but a continuum in which every perception is a vibration of the same universal field.

An interesting parallel is that both visions, albeit in different languages, deny that matter is a mere inert mechanism. In Leibniz's case, perception is distributed in discrete units; in

Eastern conceptions, perception is the very flow of reality, without interruption.

If we wanted to find a cultural bridge, we could say that pre-established harmony is, in European terms, a way of guaranteeing the unity of reality while maintaining plurality. In Eastern philosophies, unity is original and plurality is an emanation or appearance. Both models invite us to think of the universe as a living organism, but one describes it as a perfect concert of solo voices, the other as a single melody without interruption.

The relationship between Leibniz's vision and current theories on the unified field of physics.

The relationship between Leibniz's vision and current theories on the unified field of physics offers a surprising terrain for dialogue between metaphysics, Eastern philosophy, and contemporary science.

Leibniz conceives of monads as indivisible centers of perception, each with a unique perspective on the universe, but already synchronized with all the others through pre-established harmony. There is no direct exchange of information between monads; what makes their coherence possible is a fundamental, immutable principle, established at the origin by God.

In modern theories of the unified field—developed, for example, in theoretical physics from Einstein to the most recent research on string theory—the universe is described as a continuous network of interactions that manifest themselves at various levels but are ultimately expressions of a single underlying structure. In this model, space, time, matter, and energy are not independent entities but local variations of the same universal field.

Here a fascinating parallel emerges: Leibniz's pre-established harmony recalls, in philosophical terms, what physicists seek in an equation capable of encompassing all the fundamental forces . In Leibniz's thinking, God is the guarantor of synchronism; in contemporary physics, this function is attributed to mathematical coherence and the intrinsic symmetries of the field. In both cases, the universe is conceived as an orderly and deeply interconnected system.

The bridge with Eastern philosophies further enriches this comparison. Vedanta sees Brahman as absolute consciousness, the origin and substance of everything, in which differentiation is illusory. Similarly, Taoism describes the Tao as a seamless flow, the unifying principle of all phenomena. These views resemble the notion of a unified field: a continuum that sustains all forms and events, where distinctions are local manifestations of a global principle.

The substantial difference is that Leibniz maintains the irreducible plurality of monads: each point has its own perspective that never dissolves, even if it is part of the universal order. Eastern traditions, on the other hand, tend to consider plurality as an appearance destined to be reabsorbed into unity. Physics, curiously, moves between these two visions: on the one hand, it describes a single field; on the other, it recognizes that particles—like monads—possess distinct states and singular properties.

Today, some bold interpretations of quantum mechanics—such as the hypothesis of cosmic entanglement—have brought the theme of universal correlation back into vogue. In a quantum system, distant particles can manifest linked states instantaneously, without direct exchange of information, recalling Leibniz's pre-established harmony. This does not mean that physics confirms his system, but that his intuition of an invisible and total order finds a powerful echo in the most daring research in science.

Criticism and philosophical legacy — From Leibniz to Schelling.

Spinoza's thinking, so compact and rigorous, has never left anyone indifferent. At the end of the 17th century, his *Ethica* circulated in Europe as a book that was as admired as it was feared. The formula *Deus sive Natura* appeared radical: God and the universe as a single infinite substance. For some philosophers, it was a blow to the traditional conception of divinity.

Leibniz admired the geometric coherence of the Ethics, but rejected the identification of God with the totality of nature. For him, God was a personal being, endowed with will and intentions. Spinoza's impersonal and necessary vision seemed to him incompatible with divine freedom.

Leibniz developed his theory of monads precisely in tension with Spinoza. Monads were infinite simple substances, each with a unique perspective on the universe. He did not share the idea of a single substance common to everything. He believed that Spinozian monism erased individuality. In a letter to Sophia of Hanover in 1704, Leibniz wrote that Spinoza "*clouds respect for Providence by reducing it to a blind mechanism.*" Although he used harsh words, the harmonious organization of monads reversed Spinoza's unitary vision.

In the 19th century, Spinoza's name returned to prominence thanks to German Romanticism. Friedrich Wilhelm Joseph Schelling considered him the "father" of the philosophy of nature. In 1801, in Jena, Schelling explained to his students that Spinoza was the first to understand that mind and matter were two forms of the same reality. For Schelling, this was the basis for overcoming Cartesian dualism and uniting science and metaphysics.

However, Schelling transformed Spinoza's substance. His "Absolute" was not static but dynamic, capable of developing and manifesting itself in different degrees. In "*The Philosophy of* Nature," Nature is creative and in constant metamorphosis. In this sense, Schelling attempted to correct what he considered Spinoza's limitation: an immobile universe, already complete in its perfection.

The cultural link between Spinoza and Schelling can be seen in an episode from 1802. During a meeting with Hegel and Hölderlin, Schelling passionately defended Spinoza's "unity of being," arguing that without it, the correlation between organic life and physical laws could not be understood. Hegel, who had also studied Spinoza, observed that *'every philosophy must begin with* Spinozism', recognizing the foundational role of his ideas.

Thus, amid criticism and reinterpretations, Spinoza's monism survived. Leibniz challenged it by constructing a mosaic of individual substances. Schelling embraced it, but made it alive and creative. Both maintained a constant dialogue with Deus sive Natura, a sign that the rational vision of panpsychism, born in Amsterdam in the 17th century, continues to generate new philosophies.

Deus sive Natura in German scientific-romantic thought.

For Schelling, Spinoza's single substance was the matrix from which both mental processes and physical phenomena could be understood. In a lecture in Jena in 1801, he explained that a true study of nature must never separate chemistry from philosophy, because both show the different modes of the same reality. Alessandro Volta's electrochemical discovery and Humphry Davy's research on chemical reactions were interpreted by Romantic philosophers as manifestations of the 'universal organism' Spinoza spoke of.

The link was also cultural. In the rooms of Jena, Goethe discussed comparative anatomy and botany with Schelling. Goethe, who had read Spinoza in his youth, considered Deus sive Natura the key to understanding the metamorphoses of living forms. In his research on plant morphology, he saw plants as an expression of the unique life of Nature. Schelling integrated this vision with Luigi Galvani's and 's new discoveries on animal electricity, arguing that the vital principle and physical forces belonged to the same unified framework.

German scientific-romantic thought moved along two tracks: on the one hand, empirical experimentation, and on the other, the conviction that every phenomenon was part of an organic whole. This whole took the form of infinite substance in Spinoza, while for Schelling it was the evolving Absolute. Ritter's observations on light and magnetism were read as steps towards revealing the 'nerves' of Nature, a living body in which mind and matter were never separated.

In this climate, physics and chemistry became tools for confirming what philosophy had already intuited. When Davy isolated potassium and sodium in 1808, Schelling commented that each new chemical element revealed a degree of articulation of the Absolute. The single substance was not an abstract concept, but a reality that science could 'touch' through its fundamental forces.

For the German Romantics, accepting Spinoza meant rejecting the clear separation between spirit and matter. It meant reading Nature as a living book, in which the laws of crystals, photosynthesis, or the propagation of heat were different pages of the same work. Thus, the scientific-Romantic program was, in fact, a transposition into the laboratory of Deus sive Natura: a single substance, infinite manifestations, and a common effort to understand them without fragmenting their origin.

Goethe and romantic science — Organism and living nature.

Johann Wolfgang von Goethe was not only a poet. Goethe was a tireless observer of plants, animals, and natural phenomena. He traveled throughout Europe with a notebook in hand, noting shapes, variations, and colors. For Goethe, nature was not an in r inert machine. It was a living organism, endowed with an inner principle that expressed itself in every detail.

In 1784, in Ilmenau, Goethe studied the remains of a human skull. He observed a small frontal ossification and concluded that humans, like other mammals, possessed an *'intermaxillary bone'*. The discovery confirmed the morphological unity between humans and animals. The hypothesis, met with skepticism by anatomists, revealed his way of thinking: not to separate, but to understand continuity.

Goethe transferred this view to botany. During his trip to Italy in 1786-88, between Verona and Palermo, he searched for the original form of the plant. In his famous "*Metamorphosis of Plants*" (1790), he described a unifying principle. Every plant organ, from seed to flower, was a transformation of the same "primordial leaf." Here, the plant was not a mechanism, but a vital process undergoing continuous metamorphosis.

His contact with the Mediterranean profoundly influenced his thinking. In Palermo, in the botanical garden, Goethe saw tropical and Mediterranean plants developing into harmonious structures. He perceived the creative power of light and climate. He wrote that *'nature is always the manifestation of* God' (*die Natur ist Gott immerdar*). This phrase revealed his adherence to a Spinozian vision of Deus sive natura, experienced not as an abstract theorem but as a sensory experience.

Goethe's romantic science opposed Newton's mechanistic reductionism. In the debate on colors, he rejected the idea that the phenomenon was only refraction and wavelength. For Goethe, color arose from the interaction between light and darkness mediated by the human eye. Perception, therefore, was not passive. The observer participated in the physical event.

This approach influenced thinkers such as Schelling and Humboldt. Schelling saw in the philosophy of nature a reflection of Goethe's insights. Humboldt, exploring Latin America, applied the principle of organic interconnection between climate, flora, and faun . Both recognized nature as having a vital and unified dimension that could not be reduced to the sum of its parts.

Goethe did not formulate a closed theoretical system. He preferred the phenomenological method: observing without forcing reality into abstract schemes. In his study, the organism was always more than the sum of its functions. It was a living form, a cosmic rhythm, a synthesis of creative forces.

When Goethe spoke of 'living nature', he was not referring only to plants and animals. He saw the entire Earth as a being, capable of breathing and growing. It was a vision close to panpsychism: every part of nature, from stones to clouds, participated in the same universal vitality.

The philosophical and scientific influences that shaped Goethe's thinking about nature.

The main philosophical and scientific influences that shaped Goethe's thinking about nature were many and intertwined, developing throughout his life, fueled by travel, encounters, and reading.

The main influence came from Spinoza and his unified vision of being. Goethe read Spinoza in his youth and was deeply impressed by him. The idea of Deus sive Natura (God and Nature as two names for the same reality) became a vital principle for him. Not an abstract deity, separate from the world, but nature itself as a manifestation of the divine. This vision provided Goethe with the philosophical basis for conceiving of nature as a living, interconnected organism.

Renaissance naturalism and the vitalist tradition also exerted a strong influence on Goethe's thinking. He moved among botanists, physicians, and anatomists who upheld a pre-Newtonian tradition of science, in which life was considered an active and creative principle. Goethe studied authors such as

Paracelsus and van Helmont, who spoke of invisible vital and l forces and correspondences between microcosm and macrocosm. This background fueled his belief that forms and functions were expressions of an internal dynamic, not just mechanical causes.

It is undeniable that Goethe's encounter with scientific Enlightenment influenced his thinking. While criticizing reductionism, Goethe learned from the great scientists of his time. Linnaeus' research on the classification of species influenced him, but Goethe added the transformative and historical dimension of living forms. Linnaeus saw fixed categories; Goethe saw change and metamorphosis.

During his trip to Italy, direct contact with the Mediterranean landscape, the botanical gardens of Palermo, and the light and climate of southern Italy reinforced his perception of nature as a creative force. His Italian experience led him to formulate the theory of the "primordial leaf" in "The *Metamorphosis of Plants*."

The debate with Newton and optical science also played an important role. Opposition to Newton's theory of light was an act of intellectual independence. Goethe rejected a view in which perception was passive and reduced to physical measurements. Influenced by direct observational methodology, he proposed a science capable of integrating subject and object, phenomenon and observer.

His dialogue with Friedrich Schelling, who developed a philosophy of nature in which matter and spirit are two aspects of the same reality, reinforced Goethe's idea of vital interconnection. Alexander von Humboldt, explorer and naturalist, with his early ecological conception, offered empirical confirmation of Goethe's organic vision.

Finally, the discovery *of the 'os* intermaxillare' in human anatomy, also identified in mammals, was for Goethe tangible proof that life expresses itself in common structures. This type of observation reinforced his idea that nature resorts to shared models of form and adapts them.

Goethe, therefore, was not trained in a single school of thought. He intertwined philosophy, experimental science, direct observation, and poetic sensibility. The result was a conception of nature as a living, creative, and spiritually significant entity, balanced between observational rigor and metaphysical intuition.

Goethe's concept of 'Urpflanze' and modern theories of evolution.

Goethe's concept of *Urpflanze* (the "*original plant*") is more closely linked to modern theories of evolution and biology than might appear at first glance, despite originating in a historical context prior to Darwin's formulation.

For Goethe, the Urpflanze was not a real species, but an ideal model: a fundamental morphological structure capable of transforming itself into all other plant forms. In "*The Metamorphosis of Plants*" (1790), he described how every plant organ—root, leaf, flower—derived from variations and adaptations of the same primordial form, the "*archetypal leaf.*" This view emphasized continuity and transformation, concepts that are now at the heart of evolutionary biology.

Modern theories of evolution, based on the principle of common descent, assert that all species derive from shared ancestors. Genetic research confirms that conserved patterns and sequences exist at the molecular level, just as Goethe intuited by observing morphological analogies. Phylogenetic studies, which reconstruct plant family trees, demonstrate in scientific terms what Goethe perceived on a phenomenological level: variety arises from a common basic model, shaped by time and adaptation to environments.

Goethe, influenced by naturalism and partly by vitalism, interpreted plant form as a dynamic process that was more h l than static. Today, this idea resonates with concepts such as phenotypic plasticity and modular development, which explain how a single genetic structure can produce different forms in response to environmental conditions.

A direct parallel can also be found in evolutionary developmental biology (*evo-devo*), which studies the mechanisms through which small variations in regulatory genes modify the architecture of an organism. The concept of the 'primordial leaf' can be seen as a pre-scientific metaphor for these structural modules: elements that repeat and adapt, giving rise to the diversity of the plant kingdom.

In short, Goethe's Urpflanze anticipates—in poetic and observational language—the core of what modern science describes as common ancestry, adaptive variation, and morphological transformations. The difference is that Goethe attributed an internal life force to this process, while contemporary biology seeks its causes in genetic and ecological mechanisms.

Schelling and the nature-spirit identity. Philosophy of nature.

Friedrich Wilhelm Joseph Schelling was in his early twenties when he began to talk about a bold idea: nature and spirit are two manifestations of the same reality. Not two separate worlds, but two faces of a single substance. The year is 1797. In Jena, Schelling frequents Goethe, Schiller, and the young Hegel. The atmosphere is electric. Romanticism is emerging, and the Enlightenment is losing its brilliant mechanistic certainty.

Schelling observed nature not as an object to be studied, but as a living subject. He wrote that "*nature is visible spirit, spirit is invisible nature*." This sentence encapsulates the heart of his philosophy: the universe is not made up of matter alone, nor of thought alone, but is a living being that thinks through us and manifests itself in phenomena.

During his years in Würzburg and then Munich, Schelling studied geology, botany, and the then-emerging field of physics.

He saw in polarity—magnetism and electricity—a universal principle connecting physical and vital processes. Goethe was fascinated by this, recognizing in these ideas a harmony with his own organic view of nature. Alexander von Humboldt found in them a language to describe the interdependence of climates, lands, and living forms.

Schelling's philosophy of nature is also an aesthetic language. For him, the artist does not invent. The artist makes the internal creativity of nature visible. In a letter to Hegel in 1802, Schelling wrote that poetry and science must be allies: the former grasps totality in sensory perception, the latter recognizes it in conceptual formulation.

A significant episode took place in 1800, when Schelling published "*Erster Entwurf eines Systems der Naturphilosophie*" (First Draft of a System of Natural Philosophy). He describes nature as an organism evolving towards consciousness. Each degree of organization—minerals, plants, animals—is a stage in a single process. This vision, although predating Darwin, shares with evolutionary biology the idea of gradual development and continuity of life.

The following years saw him in Berlin, where his influence also extended to Romantic painting. Caspar David Friedrich's landscape, with its forests shrouded in mist, is spiritually akin to Schelling's philosophy. It is not just nature represented, but nature that thinks.

The concept of nature-spirit identity is close to panpsychism. Schelling, while not using this term, attributes an inner dimension to matter. It is not an inert aggregate. It is endowed with a principle of self-organization and implicit perception. Every form, according to him, is consciousness at the stage appropriate to its level of complexity.

From this perspective, human thought does not arise from nothing. It is the highest point of a process that nature has always carried out. The body and mind are the same phenomenon seen from two sides. Spinoza's 'Deus sive natura' thus enters a new phase: not only are God and nature one and the same, but this unique reality evolves, feels, and creates.

'Natura naturans' and 'Natura naturata' in Schelling.

In Schelling, the pairing *of 'natura* naturans' and *'natura* naturata' is much more than a reference to Spinozian terminology. It is the beating heart of his philosophy of nature, a vision in which the dynamic and creative element of reality is intertwined with its concrete manifestation.

For Schelling, '*natura* naturans' is nature that acts, creates, invents. It is not a simple set of mechanical laws, but a living and generative force, capable of producing ever new forms. It is the invisible and spiritual side of nature, that which is expressed in processes but cannot be reduced to any of their individual expressions. It is pure movement, creative impulse, a sort of "cosmic consciousness" that operates even when we do not see it.

"*Natura* naturata," on the other hand, is the result of this activity: the world of fixed forms, organisms, landscapes, organized matter. It is "accomplished" nature that we can observe, measure, and describe. It is stabilized reality that bears the imprint of the creative force but does not coincide with it.

Schelling sees this relationship as a constant dialogue. "*Natura* naturans" transforms into "*natura naturata*," and "*natura naturata*" still contains within itself the seed of creativity that generated it. This means that every mountain, every living body, every galaxy is not only a product, but a fragment of the universal creative force.

For Schelling, this vision breaks down the separation between matter and spirit. Spirit is not an external principle that shapes matter: it is matter itself as conscious, 1 in its highest form. Hence his identification of nature and spirit: two sides of the same processual reality.

The link with dynamic and spiritual nature is evident. Schelling does not describe a static universe, but a living organism in evolution. The physical world ("*natura* naturata") is not an end point. It is part of a continuum in which the spiritual dimension ("*natura naturans*") is always active, ready to generate new forms and new expressions of life.

This perspective profoundly influenced German Romanticism. Goethe, for example, saw in the metamorphosis of plants a concrete case of continuous transition between creative force and completed form. Humboldt, exploring the Americas, interpreted ecological interdependence as a direct manifestation of that "invisible force" that unites natural systems.

If we look at contemporary discussions on panpsychism or diffuse consciousness, Schelling's idea remains surprisingly relevant. "*Natura* naturans" could be read as the informational or conscious principle of the universe; "*natura* naturata" as its translation into stable configurations, from elementary particles to biological structures.

Comparison on the question of nature and spirit.

The comparison between Schelling and Hegel on the question of nature and spirit reveals two approaches that are as close in their roots as they are divergent in their development.

Schelling, as we have discussed, conceives of nature and spirit as two manifestations of the same reality. There is no clear hierarchy: matter is not a lower step to be overcome, but is already imbued with spirituality. "*Nature* naturans" is the creative and invisible principle, "*nature* naturata" is its sensible manifestation, but both belong to a continuous cycle. In this framework, spirit is not born by separating itself from nature: it is nature itself at its highest degree of consciousness.

Hegel, on the other hand, while starting from common ground, develops a more dialectical and hierarchical vision. For him, spirit unfolds through a necessary sequence of stages, from the most primitive moment to self-conscious freedom.

An example clarifies the difference: for Schelling, an alpine landscape, with its geological forces and plant growth, is a living spirit that determines itself in mineral and biological forms. For Hegel, the same landscape is indeed part of the Absolute, but it remains a necessary and incomplete moment, surpassed in the movement towards higher forms of self-consciousness, which are fulfilled in human activity, politics, art, and philosophy.

It could be said that Schelling favors an organic and cosmic vision, while Hegel favors a historical-rational vision. The former insists on vital continuity, the latter on logical progression. Both seek the unity of nature and spirit, but Schelling looks to life as the key to interpretation, Hegel to reason.

This divergence had consequences for subsequent thought: Schelling opened up avenues towards the idea of a consciousness diffused even in matter, influencing romantic movements and, indirectly, some forms of modern panpsychism. Hegel directed attention towards the role of history and institutions in shaping the spirit, fueling the political and ideological debate of the 19th and 20th centuries.

Phenomenology and diffuse consciousness. Husserl, Merleau-Ponty

When Edmund Husserl published his Logical *Investigations* in 1900, he did so in a Europe still strongly marked by scientific positivism. Psychology tended to reduce consciousness to a set of neurophysiological processes. Husserl made a radical move: he brought consciousness back to the center of philosophical inquiry, not as an "object" among others, but as the original field in which everything appears.

The phenomenology he inaugurated is not limited to studying the contents of the mind. It suspends judgment on the existence of things—the famous *epoché*—in order to investigate how the world manifests itself to consciousness. Husserl speaks of the "world of life" (*Lebenswelt*), that horizon of everyday experience that precedes any theory or measurement. This world, according to him, is already endowed with meaning, because consciousness does not passively receive impressions: it structures, orders, and connects.

In 1933, Husserl, now professor emeritus at Freiburg, encountered the dramatic intellectual closures of his era, the purge of Jews, and the ban on access to libraries. However, Husserl's lectures at the Sorbonne in 1929 had already attracted a young French philosopher, Maurice Merleau-Ponty. He would take phenomenology beyond the descriptive dimension, towards a bodily immersion in the world.

In 1945, Merleau-Ponty published *Phenomenology of Perception.* It was the post-war period. Paris was experiencing a climate of cultural reconstruction. The philosopher introduced a decisive idea: the body is not an object among objects, but the means through which we are in the world. Consciousness does not 'reside' in the head, but spreads through gesture, gaze, and action. The hand that grasps, the eye that explores, the step that measures the road are not peripheral activities: they are consciousness in action.

Hence the notion of diffuse consciousness: not confined to a mental center, but distributed throughout the entire relationship between the subject and the environment. A pianist improvising, a farmer reading the signs of the weather in the sky, a child learning to walk... all experience a bodily knowledge that does not pass through discursive thought, but is nonetheless sensitive intelligence.

The link with panpsychism is subtle but profound. Husserl and Merleau-Ponty describe a field of experience where the world s never "external" to consciousness. If consciousness is diffuse and embodied, one can think of a continuity of perception and intentionality even in non-human living forms. From this perspective, minerals, plants, and animals are not mere objects: they have their own ways of being-in-the-world.

Merleau-Ponty, fascinated by the biological sciences, often cites the physiology of visual perception, studies on octopuses, and the response of plants to stimuli. In a note from 1959, he writes that "Nature is at the origin" and that "we are its thinking body." This sentence encapsulates the idea: humanity is not above nature, but is one of its forms of self-awareness.

Husserl's walks in the gardens of Freiburg, recounted by some of his students, are eloquent images. He did not look at a tree as a botanical object, but as a living phenomenon in its entire sensory entirety: shadow and light, space and time perceived together. Merleau-Ponty would have recognized in this gaze the same unity that binds the root of a plant to the earth and to atmospheric phenomena: a system of meanings that are nonverbal but real.

In this sense, phenomenology becomes a bridge between modern thought and ancient questions about "Deus sive natura." If consciousness is diffusion in the world and the world is a web of senses, perhaps reality itself—visible and invisible—is a network of intertwining consciousnesses. Not an isolated man contemplating nature, but nature contemplating itself through man.

PHENOMENOLOGY AND DIFFUSE CONSCIOUSNESS. MERLEAU-PONTY.

Maurice Merleau-Ponty takes phenomenology in a decidedly embodied direction, shifting the focus from consciousness understood as purely mental activity to consciousness experienced through the body. For him, the body is not an object among objects, but the point of origin of all perception and action.

In "*Phenomenology of Perception*," written in a Paris just emerging from war, Merleau-Ponty states that we are body before we are reflective mind. Vision, touch, and movement are not simply biological functions: they are ways in which consciousness opens up to the world. A gesture—such as lifting a cup or entering a room—is not the execution of a mental program, but an act in which meaning and movement coincide.

This perspective generates the notion of diffuse consciousness: consciousness does not live in a "control unit" hidden in the brain, but is distributed in the relationships between the body and its surroundings. For example, when improvising, a pianist does not think note by note. Their body "knows" where to go, anticipating and responding with a silent knowledge that is already consciousness.

Merleau-Ponty observes that this relationship is bidirectional: the world is not a silent stage on which man moves, but an active interlocutor. A rock, a forest, the water of a river influence the possibilities of action and perceptual patterns. When the body moves, it dialogues with these presences, adapting and modifying itself.

The link with panpsychism manifests itself precisely here: if consciousness is a web of embodied relationships, then the 'life of consciousness' is not exclusive to human beings. Every organism can have its own way of perceiving and inhabiting the world. The balance of an animal moving in a complex environment, or the reactivity of a plant to light, become manifestations of a primary intentionality.

Merleau-Ponty, in his notes from the 1950s, writes: "*Nature is our oldest body*." This sentence sums up his vision: we are nature that thinks and perceives itself. The human body, like that of every other living form, is a node in an immense perceptual fabric that includes the entire environment.

In recent decades, the phenomenological insights of Husserl and Merleau-Ponty have found resonance in unexpected directions, even at the heart of the physical sciences. So-called "quantum consciousness" (theories proposed by physicists and neuroscientists such as Roger Penrose and Stuart Hameroff) suggests that mental processes could emerge from phenomena of quantum coherence within neuronal structures, such as microtubules.

In this perspective, consciousness would not be a mechanical by-product of the brain, but a fundamental interaction with the very laws of physical reality. The boundary between mind and matter would become blurred, not unlike what Merleau-Ponty intuited when he described the body as a "node" in a web of relationships that binds the living to the world. If consciousness is diffused throughout the body and the body is immersed in nature, then there could be a universal perceptual substrate, compatible with the panpsychist hypothesis that every entity participates in some form of experience.

The new contemporary panpsychism, supported by philosophers such as Philip Goff, takes up this idea in a modern key: if consciousness is an intrinsic property of matter, it manifests itself in varying degrees, from the simplest molecular lattice to complex human self-perception. In this sense, Husserl and Merleau-Ponty, while speaking in phenomenological language, would have recognized quantum physics as an unexpected cultural ally, a field capable of giving their vision a naturalistic foundation.

Today, the encounter between embodied phenomenology and quantum science opens up a horizon in which Spinoza's "Deus sive natura" is reinterpreted in cosmological terms: consciousness, rather than being the privilege of a few beings, becomes an invisible web that runs through the cosmos. Man, with his thinking body, would be one of the ways in which that universal consciousness takes on words, gestures, and memory.

Bergson and the "élan vital." Time and life as cosmic consciousness.

Henri Bergson entered the philosophical scene at the end of the 19th century as an author capable of speaking to Parisian literary salons as well as scientific laboratories. Born in 1859, he formed his thinking in an era dominated by positivist confidence in deterministic laws. But from his earliest writings, such as *'Essay on the Immediate Data of* Consciousness' (1889), his desire to restore to life a breath that escapes the rigid patterns of mechanics emerges.

Bergson observed nature as movement, not as a collection of static states. In Creative *Evolution* (1907), he introduced the concept of "*élan vital*": an original, creative force that runs through living matter and emancipates it from pure inertia. This force is not a physical law, but a universal tendency that is embodied in biological processes, human gestures, and even historical transformations. For Bergson, life invents, like an artist who never repeats the same painting.

In Paris, during the years when Bergson taught at *the Collège de France*, his lectures attracted mixed crowds of philosophers, poets, scientists, and politicians. Raffaello Franchini recounts that even Proust, before writing *In Search of* Lost *Time*, read Bergson to understand how memory is not a fixed archive, but a flow that selects and reinvents. The writer found in Bergson the conceptual support for inner time, the "*temps retrouvé*."

Bergson distinguishes between scientific time, made up of measurable moments, and real duration (*durée*), experienced by consciousness as a continuous, qualitative flow. In his writings, experienced time behaves like music: a melody that exists only as movement, where each note derives meaning from what precedes and follows it.

This vision naturally opens up a link with panpsychism. If life is a current that flows through matter, and if consciousness is the condition for experiencing duration, it is possible to think that consciousness is not a late privilege of man. In simpler forms, it could be present at all levels of being, like a thin thread connecting the algae that bends to the current, the animal that builds a shelter, and the human being who composes a symphony.

During the 1927 Nobel Prize in Literature award ceremony for Bergson, the speakers emphasized how he had given philosophy "*the poetic daring of science and the conceptual precision of art*." This synthesis explains why his influence crosses disciplinary boundaries and historical decades. At a time when quantum physics was beginning to challenge the idea of a mechanical universe, the "*élan* vital" appeared as an intuitive anticipation of a creative and unpredictable reality.

In his last lectures, Bergson admits that the "*élan vital*" cannot be defined by formulas, because it is "closer to music than to geometry." When read today, this metaphor surprisingly resonates with contemporary theories of diffuse consciousness and even with some quantum hypotheses: life as a wave that cannot be reduced to fixed points, but vibrates in every form, from the seed that germinates to the thought that is born.

VIII. The concept of consciousness.

Consciousness is perhaps the most fascinating and elusive mystery of philosophy and science. It is not only the ability to perceive the world, but also to perceive oneself while experiencing it. It is the internal theater in which sensations, memories, and thoughts flow, and at the same time the stage on which the images of what we call "reality" are projected.

Philosophers have been questioning it for centuries. Plato sees it as illuminated by the idea of Good, Descartes links it to the "cogito" — the thought that proves the existence of the subject. In the 20th century, Thomas Nagel wrote the famous essay "*What is it like to be a bat?*" (1974), to show how every experience has a subjective character that cannot be reduced to purely physical measurements. Twenty years later, David Chalmers defined this enigma as "*the hard problem*": explaining why brain processes produce inner experiences and not just mechanical reactions.

Neuroscience offers new tools and languages. Using *brain imaging* techniques, researchers map the areas that are active when a person remembers, evaluates, or dreams. Some models, such as Bernard Baars' "*Global Workspace Theory*," compare consciousness to a "central square" in the brain where information is shared between different systems. Others, such as Giulio Tononi's "*Integrated Information Theory*," propose measuring consciousness as the degree of integration and complexity of information.

But consciousness is not just a function. In today's debate, a distinction is made between phenomenal consciousness, the direct and qualitative feeling of experiences, and access consciousness, the use of that information to reason, speak, or act. It is the difference between savoring the taste of coffee and describing it in words.

Alongside neuroscientific models, there is a more speculative current. Some thinkers imagine consciousness as a unified field, similar to a pervasive physical phenomenon. In this framework, panpsychism proposes that every fragment of matter has an elementary form of experience. These ideas echo, in modern language, the visions of Spinoza, Leibniz, and Bergson: nature as a living fabric in which nothing is truly devoid of perception.

In these pages, the concept of consciousness will be addressed from multiple angles: operational definitions, theoretical distinctions, big philosophical questions, integrative hypotheses, and ontological perspectives. From the laboratory to the living room, from Eastern meditation to astrophysics, the theme crosses boundaries and disciplines. As William James wrote in 1890, "*consciousness is a river*," and every age attempts to follow its current, without ever capturing its entire source.

Operational definitions of consciousness. Philosophy and neuroscience.

Defining consciousness means walking on ground that changes as you walk on it. For a philosopher, it is the core of subjective experience. For a neuroscientist, it is a set of integrated processes that make it possible to perceive, think, and act. The challenge is to find a definition that works both in the language of ideas and in that of measurements.

In 1890, William James described it as "*a stream of consciousness*" that flows without interruption. Centuries earlier, Aristotle considered it to be the ability to perceive both the object and oneself in the act of perceiving it. Spinoza included it in his grand monistic scheme: mind and body are two modes of the same substance, Deus sive natura.

In the modern laboratory, consciousness is operationally defined as the state in which an organism is awake, aware of the outside world, and able to respond coherently. This definition helps doctors distinguish between a person in a coma and an alert person. It helps anesthesiologists identify the boundary between sedation and perception.

Neuroscience needs precise criteria. In the 1990s, Bernard Baars proposed the "*Global Workspace Theory*": consciousness is a central "workspace" where information from different brain systems becomes available to the whole brain. Metaphorically, it is like a lit stage surrounded by dark wings, where only what enters the light can be used by actors and directors.

Giulio Tononi, with his "*Integrated Information Theory,*" measures consciousness based on the amount of information integrated into a system. An awake human brain has a very high value. A computer, even though it processes enormous amounts of data, does not have the same degree of phenomenal integration. This definition allows for experiments on patients in minimally conscious states and even discussions about the degree of consciousness in animals.

However, the operational definition does not exhaust the philosophical question. Thomas Metzinger, philosopher and neuroscientist, argues that consciousness is a model of the self constructed by the brain—useful for navigating the world, but devoid of metaphysical substance. Others, such as Evan Thompson, see consciousness as a relational dynamic between organism and environment, in continuity with Buddhist traditions.

Each definition is partial. When Alan Turing wrote his 1950 essay "*Computing Machinery and Intelligence*," he wondered whether an artificial system could have consciousness or only imitate conscious behavior. The question remains open. Philosophy insists on the qualitative meaning of experience, science on the mapping of its neural correlates.

Ultimately, the operational definition of consciousness is like a map of a territory: necessary for orientation, but incapable of rendering the landscape in its entirety. Consciousness, the real thing, is not just in the definition. It is in the life we perceive as we try to define it.

The main methodological differences in the approach to consciousness between philosophy and neuroscience.

The main methodological differences between philosophy and neuroscience in the study of consciousness arise from two different ways of understanding and investigating the phenomenon.

Philosophy moves in a conceptual and often speculative territory. The philosopher starts from subjective experience—what one feels, what appears to the mind—and seeks to clarify its meanings, implications, and possible structures. The approach is analytical on the level of ideas, not experimental data.

William James describes consciousness as a continuous "stream," emphasizing the quality of experience rather than measurement.

Thomas Metzinger constructs theoretical models of the conscious self to explain why we perceive a unified "I."

Evan Thompson, in dialogue with contemplative traditions, studies consciousness as a dynamic relationship between organism and environment.

Philosophy uses methods such as conceptual analysis, argumentative thinking, and historical comparison, and often questions issues that do not yet have an empirical answer: for example, why neural activity "becomes" lived experience (*the hard problem*, as Chalmers defines it).

Neuroscience, on the other hand, treats consciousness as an observable and measurable phenomenon. Scientists start from clear operational criteria—for example, a state of wakefulness and responsiveness—and seek to identify the neural correlates of these states.

Methods include *brain imaging*, neural stimulation, behavioral studies, and computational neuroscience. The goal is to establish a verifiable link between brain functions and conscious states.

In summary, philosophy explores the "what" and "why" of consciousness, questioning the nature of experience and its ontological implications. Neuroscience explores the "how," searching for the physical and biological mechanisms that allow consciousness to emerge.

The dialogue between the two is essential: without the empirical precision of neuroscience, philosophy risks remaining suspended in the abstract; without the radical questions of philosophy, science risks reducing consciousness to mere data, losing sight of the complexity of experience.

Key figures in the field of consciousness.

Here is a comparative summary in table form, comparing key figures in philosophy and neuroscience, highlighting their respective operational definitions of consciousness and their impact on the study of the topic.

Table14- Key figures in philosophy and neuroscience for the development of the theme of consciousness.

Field	**Key figure**	**Operational definition of consciousness**	**Impact on the study of consciousness**
Philoso phy	Aristotle (4th century BC)	The ability to perceive an object and, simultaneously, perceive oneself in the act of perception.	He established the distinction between simple perception and self-awareness, influencing the philosophy of mind for centuries.
Philoso phy	Spinoza (17th century)	Mind and body as two modes of the same substance (*Deus sive natura*), consciousness as an attribute of the single substance.	He supported monism, paving the way for integrated and non-dualistic views of consciousness.
Philoso phy	William James (1890)	"*Stream of consciousness*," continuous and unified experience.	He introduced a dynamic perspective, influencing psychology, phenomenology, and modern fiction.
Philoso phy	Thomas Metzinger (21st century)	Model of the self generated by the brain, devoid of autonomous metaphysical reality.	He offered a theoretical bridge between philosophy and neuroscience, guiding studies on the self and the illusion of the ego.

Field	Key figure	Operational definition of consciousness	Impact on the study of consciousness
Philoso phy	Evan Thompso n (21st century)	Consciousness as a relational process between organism and environment.	Integrates science and contemplative practices, expanding the field of neuroscience toward intercultural neuroscience.
Neuros cience	Bernard Baars (1990s)	Global Workspace Theory: consciousness as a "workspace" where information becomes globally accessible.	Provided a testable framework for cognitive experiments and artificial intelligence models.
Neuros cience	Giulio Tononi (21st century)	Integrated Information Theory: degree of consciousness measured by the amount of information integrated into a system.	Crucial in clinical settings for assessing altered states of consciousness, and in the debate on animal and artificial consciousness.

There are several cases in which empirical advances in neuroscience have prompted philosophers to reformulate their definitions of consciousness, just as philosophical ideas have guided new strands of neuroscientific research. Here are some concrete examples of this two-way exchange:

FROM NEUROSCIENCE TO PHILOSOPHY.

Discovery of the neural correlates of consciousness (NCC).

Thanks to techniques such as fMRI and electroencephalography, scientists have identified brain networks—such as the "*default mode network*"—that are active during self-reflection and spontaneous thinking. This evidence has led philosophers such as Evan Thompson and other contemporary phenomenologists to integrate into their models the notion that consciousness is not only an 'event' but also a 'background state' structured by continuous neural processes.

Integrated Information Theory (Giulio Tononi)

The proposal to quantify consciousness through the degree of information integration has stimulated philosophers of mind to redefine consciousness not as an "indivisible" quality, but as a scalar property. This idea has renewed classic discussions about the "degree" of consciousness in animals, children, and even machines.

FROM PHILOSOPHY TO NEUROSCIENCE.

Stream of consciousness (William James).

James' idea of the continuity of experience has inspired neuroscientific experiments aimed at understanding how the brain processes information in real time and how it maintains coherence in the "mental movie" despite rapid perceptual changes. This has resulted in studies on temporal *binding* and rapid synaptic plasticity.

Model of the self (Thomas Metzinger).

The thesis that the 'self' is a useful but illusory neural model has prompted researchers to investigate the neural correlates of the sense of identity, leading to studies on disorders such as depersonalization and self disorders in deep meditation.

"Global Workspace Theory" (Bernard Baars) and philosophical roots.

Although formulated in the cognitive-scientific field, this theory is inspired by philosophical questions about the unity of experience. This has led to experiments that measure synchronization between brain areas during conscious tasks, combining empiricism and philosophical concepts of the "mental stage."

CURRENT MUTUAL DIALOGUE

Today, philosophical definitions such as "*phenomenal consciousness*" vs. "*accessible consciousness*" (Block) are used in experimental design to distinguish what to measure.

At the same time, new neuroscientific discoveries about minimal states of consciousness (coma, vegetative state, anesthesia) are leading philosophers to rethink the very concept of "experience" and its boundaries.

Phenomenal consciousness and access consciousness.

When talking about consciousness, everyday language tends to confuse what we feel with what we know we feel. Two key concepts help to distinguish these dimensions: phenomenal consciousness and access consciousness.

The first concerns subjective experience, how it feels to be in a certain state. Thomas Nagel made it famous with the question: "*What is it like to be a bat*?" (1974). He did not want to explain the biology of bats, but rather the ineffable flavor of subjective experience, which is different from any objective description. Phenomenal consciousness lives in the realm of the colors we see, the music we hear, the melancholy we feel on an autumn afternoon in Vienna, with the leaves of the Prater falling slowly.

Access consciousness, on the other hand, is more pragmatic. It deals with the information that the mind can put "on the table" in order to reason, plan, and respond. David Chalmers describes it as the ability of a system to use mental content to guide thought and action. It is what happens when we read a sentence and keep it in mind to respond, when we transform a memory into an argument in a debate, or, , when a violinist coordinates the movements of their hands and fingers to perform a piece by Bach.

The distinction between the two notions is not a textbook exercise. It has concrete consequences. A patient under general anesthesia loses both forms of consciousness. But those who dream retain phenomenal consciousness and lose most of their access consciousness: in lucid dreaming, in fact, a fragment of access is recovered, and one becomes both spectator and director at the same time.

In 1995, Chalmers made famous the so-called hard problem of consciousness: explaining how and why phenomenal consciousness emerges from physical processing in the brain. Neuroscience elegantly shows how access consciousness works—rules, neural networks, signals that enter and are shared—but when it comes to phenomenology, science becomes timid. How can we describe the bittersweet taste of a bar of dark chocolate enjoyed in a café in Turin, even if we know every synapse that is activated?

Some scholars, such as Ned Block, have insisted that it is essential to maintain the distinction. Block noted that phenomenal consciousness can exist without access consciousness. A striking example is subliminal perception: the image of a face appears for a few milliseconds and vanishes, leaving an impression that we cannot use deliberately, but which nevertheless affects our mood.

In recent times, this debate has also entered the field of experimental neuroscience. Studies with brain-damaged patients have revealed cases in which brain activity compatible with subjective experiences exists, even when the subject cannot report them verbally. This is the terrain of diagnoses of vegetative states and "minimal consciousness," where the distinction between phenomenal and access becomes crucial not only for theory but also for ethical decisions.

Culture has metabolized the distinction in surprising ways. In Virginia Woolf's fiction, stream of consciousness explores the purest phenomenal dimension. In contrast , in detective novels, the detective's mind works in access mode: it categorizes evidence, develops hypotheses, and plans strategies. Two styles of storytelling, two ways of describing the same mind.

The challenge today is to understand whether these two forms of consciousness are roots of a single tree or separate trunks that share only the soil. Philosophers debate. Scientists measure. Artists narrate. And the experience—phenomenal or access—continues to elude definitive definitions, keeping the fascination of mystery alive.

Phenomenal consciousness and access consciousness in dreams or altered states of consciousness.

The distinction between phenomenal consciousness and access consciousness becomes particularly evident when analyzing dreams and various altered states of consciousness, because in these contexts the two forms do not necessarily move in sync.

Let's talk first about ordinary dreams.

During REM sleep, phenomenal consciousness remains vivid: images, sounds, tactile sensations, and even smells are perceived with an intensity that sometimes exceeds that of wakefulness. The mind creates complex scenarios, unreal landscapes, or distorted memories. It is a rich tapestry of subjective experiences, often accompanied by strong emotions—such as the anguish of a nightmare or the serenity of flying over a beach at sunset.

Access consciousness, on the other hand, is reduced. In ordinary dreams, we do not have full control over the experience, nor are we aware that we are in a dream state. The information produced by these experiences is not generally 'brought' into the deliberative system of the mind: we cannot consciously decide to change the plot, analyze the logic of events, or interpret a sign as we do when awake.

Let's now move on to lucid dreaming. Here the situation changes. When a dreamer becomes aware that they are dreaming, access consciousness partially re-emerges: it is possible to reason, make decisions, and modify events. An experiment conducted in the 1980s by Stephen LaBerge at Stanford University showed that lucid dreamers could send predefined signals (eye movements) to observers in the laboratory, confirming that they retained some cognitive control while remaining in the dream.

Phenomenal consciousness remains intense, but is now enriched with a meta-layer of awareness, in which the subject can say to themselves: "I *am in a dream and I can act within it*." It is like going from being unconscious actors to becoming directors of a scene that takes place in a mental space.

We can also evaluate what happens in induced altered states. In contexts such as deep meditation, hypnotic trance, or the use of psychedelic substances, phenomenal consciousness can expand or change radically. More vivid colors, temporal distortions, feelings of unity with the environment. Here, access consciousness often undergoes a reduction or transforms its logic: sequential thinking may weaken, leaving room for analogical, associative connections, sometimes confusing but perceived as deeply meaningful.

For example, in the psychedelic trips described by authors such as Aldous Huxley, the subject experiences an amplified phenomenal experience but may struggle to translate it into rational sequences or communicate it as it happens.

Finally, let us examine the case of anesthesia and states of minimal consciousness.

In the case of general anesthesia, both forms are greatly attenuated to the point of almost disappearing. The case of patients in minimally conscious states is different, where neurological studies—such as those conducted by Adrian Owen's group—have shown brain activity compatible with pheno l experiences, even when the person has no conscious access to report or use them.

In summary, dreams and altered states reveal how phenomenal consciousness and access consciousness are two distinct sides of our mental life: the former can flourish even when the latter is absent or reduced, while the latter can, in certain contexts such as lucid dreaming, modulate and even direct the plot of the experience.

Connecting theories of phenomenal and access consciousness with ancient Eastern philosophies.

Modern distinctions between phenomenal consciousness and access consciousness find surprising connections with ancient Eastern traditions, such as Buddhism and Yoga, even though the language and conceptual framework are very different.

In Buddhism, particularly in the Zen and Theravāda traditions, direct experience—what we might approximate as phenomenal consciousness—is explored with great precision. The practice of vipassanā ("deep seeing") aims to cultivate continuous mindfulness, observing sensations, thoughts, and internal states as they arise, without judgment. Here, the focus is on "watching the movie of experience" in real time. It is a refinement of the phenomenology of consciousness.

In Buddhist thought, access consciousness comes into play when phenomenal awareness is used to generate insight (*prajñā, wisdom*). It is the transition from immediate perception to the ability to reinterpret and redirect experience toward liberation from suffering.

In practice, the monk who meditates on the sensation of warmth in the sun does not merely experience it: he can access that experience to reflect on the transitory nature of all things (impermanence), a deliberative use of mental content.

Raja Yoga, as formulated by Patañjali in his Yoga Sūtra (circa 2nd century AD), describes a sequence of mental states (samādhi) that can be read, with caution, in the light of modern distinctions.

In the simplest levels of meditation (*dhyāna*), the practitioner cultivates a stable phenomenal experience, free from gross distractions.

With refinement, however, the content of the experience becomes the object of mental mastery: the mind not only 'feels' but also 'directs' and 'uses' the experience. This transition recalls access consciousness: the yogi is not only immersed in inner silence, but can recognize and shape the experience at will.

Eastern philosophies do not speak of 'phenomenal' and 'access' in neuroscientific terms, but the practices they propose seem to implicitly recognise the distinction.

The Phenomenal can be defined as pure presence, "savoring" what is happening in the moment, without processing.

Access consciousness, on the other hand, can be defined as the use of awareness to control attention, direct thought, or deduce teachings.

The difference is that, in Eastern traditions, the two dimensions are often integrated into a single path: consciousness is cultivated both in its subjective intensity and in its transformative utility, with the ultimate goal of transcending the ordinary limitations of the mind.

Today, some researchers (such as Francisco Varela, a neurophenomenologist) have attempted to unite these perspectives. Studies of experienced meditators show that certain practices can increase the operational distinction between immediate experience and the ability to use it deliberately, just as the phenomenal/access framework suggests.

Mindfulness meditation, reintroduced in a clinical context, works on the phenomenology of experience (moment-to-moment awareness) but also on the functional recovery of mental content- s for therapeutic purposes, a clear bridge between the two worlds.

The following comparative table relates states of consciousness in Buddhist and yogic traditions to the concepts of phenomenal consciousness and access consciousness, highlighting similarities, differences, and points of contact for contemporary research.

The table is enriched with historical and textual links that show how, even in ancient times, Buddhism and Yoga described, with different languages and purposes, dynamics that we can now interpret in light of the distinction between phenomenal consciousness and access consciousness.

Table15- States of consciousness in Buddhist and yogic traditions compared with the concepts of phenomenal consciousness and access consciousness.

Tradition	**Practice/State**	**Ancient reference text**	**Phenomenal consciousness (direct experience)**	**Access consciousness (deliberative use of experience)**	**Historical-conceptual connection**
Theravāda Buddhism	*Vipassanā* (deep insight)	*Satipatthāna Sutta* (Majjhima Nikāya 10)	Real-time observation of the body, sensations, mind, and external phenomena, without reactivity.	Interpretation of these observations as manifestations of impermanence (*anicca*), suffering (*dukkha*), and non-self (*anatta*).	The text distinguishes between "contemplating what appears" and "understanding its nature," implicitly anticipating the passage from phenomenon to access.
Zen Buddhism	*Zazen* (sitting meditation)	Collections of kōans and instructions as in Dōgen's	Pure experience of breath, posture, and environmental	Use of this immediate presence to produce sudden intuition (*satori*),	Zen emphasizes the unity between direct perception and

Tradition	Practice/State	Ancient reference text	Phenomenal consciousness (direct experience)	Access consciousness (deliberative use of experience)	Historical-conceptual connection
		Shōbōgenzō	sounds, without conceptual discrimination.	transforming perception into insight into reality.	cognitive transformation, but maintains recognizable stages in terms of phenomenon /access.
Raja Yoga	*Dhyāna* and *Samādhi*	*Yoga Sūtra* of Patañjali (III, 1-3)	Conscious immersion in a mental object, until merging with it.	Intentional direction of attention, choice of object, and transition to *samādhi* with total mastery.	Patañjali distinguishes passive absorption from active mastery, implicitly conceptualizing the two phases of consciousness.
Hatha Yoga	*Prāṇāyāma* (breath control)	*Hatha Yoga Pradīpikā* (II, 2-3)	Intense sensory experience of the rhythm and depth of breath.	Use of breath to purify mind and body, preparing for deeper meditative states.	The text clearly indicates the transition from bodily perception to functional manipulation of the mental state.

Ancient Buddhism, in *the Satipatthāna Sutta,* shows a structure in which one moves from immediate perception (, "contemplating the body as body") to a deep understanding of its nature, equivalent to the leap from phenomenal consciousness to access consciousness.

In classical yoga, the *Yoga Sūtra* distinguishes between simple states of concentration and the deliberate use of these states to achieve liberation, parallel to the modern distinction. Some intermediate texts, such as the *Hatha Yoga Pradīpikā*, show the same logic applied to physiology and energy regulation.

Panpsychism and the phenomenal/access distinction.

Panpsychism holds that every fragment of reality, from the simplest matter to complex organisms, possesses a minimal form of consciousness. In this framework, the distinction between phenomenal consciousness and access consciousness takes on particular significance.

If every entity possesses an intrinsic "what it feels like," then even brainless systems (such as particles or molecules) would have a rudimentary form of subjective experience, however infinitely simple and inaccessible to human observers.

This experience would not necessarily involve any ability to process or use information: it would be pure "presence" or "sensitivity" to the surrounding reality.

In complex systems (such as the human brain), phenomenal experience is 'made available' to processing networks, becoming accessible consciousness.

From a panpsychist perspective, access is not universal: it is a property that emerges only when there is an organization sophisticated enough to integrate and use the contents of experience.

In the classical perspective, the phenomenal and access are two aspects of the same mind, but the phenomenal depends on specific neurobiological events.

In the panpsychist perspective, the phenomenal exists in minimal form everywhere, while access is a special case, the result of the evolution of systems that can connect experience and action. As a consequence, the "boundary" between the phenomenal and access is not only a cognitive level, but a cosmic gradient—from simple universal feeling to full use, as in beings with brains.

CONNECTIONS WITH ANCIENT EASTERN PHILOSOPHIES.

Buddhism and Yoga, especially in their metaphysical views, often postulate a "universal consciousness" or "pure awareness" that permeates every phenomenon.

In deep meditation, many texts describe states in which experience is pure and without content (universal phenomenal) and then moments in which this experience is integrated into insight or mastery (access).

Panpsychism can be read as the philosophical-scientific translation of this insight: all reality vibrates with a basic consciousness, but only in certain cases does it organize itself to become operational.

The table below shows how the minimal phenomenon of panpsychism can evolve, through increasing organization, to the consciousness of access in complex systems.

This table shows how, from a panpsychist perspective, phenomenal consciousness is universal but variable in complexity; the transition from "pure experience" to complex access emerges with the increase in informational organization and self-regulation capacity.

Table16- Progressive evolution of a minimal phenomenon to access consciousness through increasing degrees of organization. (From a panpsychist perspective)

Level	**Description**	**Example**	**Predominant type of consciousness**	**Evolutionary characteristics**
1 – Minimal phenomenal	Elementary subjective presence, pure "sensitivity" without processing.	Subatomic particle, photon, physical microstructure.	Only phenomenal (rudimentary).	There is no memorization or use of experience; it is simply "being."
2 – Primitive integrated phenomenal	Ability to register environmental changes without transforming them into complex actions.	Protists, single-celled life forms.	Phenomenal with micro-integration.	Early networks of elementary perception; beginning of internal organization.
3 – Oriented phenomenal	Direct perception associated with simple, automatic responses.	Insects, organisms with a simple nervous system.	Predominantly phenomenal with rudimentary access.	Stimulus → reaction; ability to direct behavior based on immediate experience.
4 – Emerging access	Subjective experience managed by processing networks that select and combine information.	Fish, reptiles, non-human mammals.	Phenomenic + basic access.	Experience enters the decision-making circuit; actions based on memories.
5 – Complex access	Deliberative use of experience in planning , analysis, and language.	Adult humans, some great apes.	Phenomenic + advanced access.	Phenomenic experience is rich, integrated, and guided by conscious

Level	**Description**	**Example**	**Predominant type of consciousness**	**Evolutionary characteristics**
				cognitive processes.
6 – Reflective/ metacognitive access	Ability to observe and modulate one's own consciousness, integrating it with abstractions and universal concepts.	Advanced meditators, theoretical systems of future advanced AI.	Deep phenomenal + metacognitive access.	The nature of experience is recognized and can be acted upon intentionally.

The hard problem of consciousness. Thomas Nagel and David Chalmers.

Thomas Nagel wrote in a now classic 1974 article that "knowing what it is like to be a bat" is an insurmountable challenge for current science. For an attentive reader, that sentence is not a play on words but the opening of a crack in the wall of reductionism. Nagel does not discuss echolocation or anatomy, but subjective experience, what philosophers call "qualia." It is "what it feels like" to be alive, not simply "how" an organism works.

Nagel, who grew up in the intellectual New York of the 1960s, soon realized that the physical explanation of the brain could not capture the essence of experience. His analogy with the bat was carefully chosen. An animal that is culturally, perceptually, and biologically distant, it lays bare the gap between objective description and inner experience. Anyone can read hundreds of articles on ultrasonic waves, but no one can ever transform this knowledge into the actual sensation of orienting oneself in the dark by means of an auditory "click."

Twenty years later, David Chalmers took this reasoning to a more systematic level. In 1995, during a conference in Arizona, he defined the task of explaining how subjective experiences emerge from physical processes as the 'hard problem of consciousness'. For Chalmers, it is not enough to describe neural circuits or information flows. Consciousness is not just a system for accessing data; it is a vivid field, colored by feelings, pain, pleasure, and ineffable nuances. Current science, while able to map brain activations with pinpoint accuracy, does not explain 'why' and ' ' these activations generate a certain flavor, a certain emotional tone.

The debate exploded. Some neuroscientists, such as Francis Crick—Nobel Prize winner for the discovery of the structure of DNA—replied that the problem was only "difficult" for now, and that advances in neuroscience would solve it. Chalmers remained firm in his position: you can describe all the mechanisms of a camera, but that does not explain what it means to look at the sky at sunset.

The 'hard problem' is therefore at the heart of every contemporary panpsychist theory. If consciousness seems to resist purely physicalist explanations, it is possible that it is a fundamental aspect of reality, like space or time. Some philosophers, inspired by Spinoza, have proposed that mind and matter are two perspectives of the same substance. The idea sounds speculative, but it has deep roots in the history of Western and Eastern thought.

In a university lecture hall in Sydney, Chalmers summarized his position to a diverse group of students: "If consciousness is indeed a primitive feature of the universe, our tools for studying it must include not only measurement, but also reflection on the observing subject." Some left the lecture feeling as if they had crossed the threshold into unknown territory.

The difficult problem of consciousness is not a technical issue. It is an invitation to question the relationship between reality and experience, what it means to "be" rather than "function." And perhaps this question, rather than the answer, is what keeps philosophy alive alongside science.

The difficult problem and quantum consciousness.

There are mysteries that resist even the most advanced technologies. Consciousness is one of them. Despite magnetic re , neural maps, and billions of data points, the question remains: why does an electrical impulse in the brain translate into a lived sensation? It is the question that Australian philosopher David Chalmers brought to the forefront of the debate in the 1990s with his famous distinction between the "easy problems" and the "hard problem" of consciousness.

For Chalmers, the 'easy problems' concern functioning: how we see, how we recognize faces, how we store memories. They can be addressed with neuroscientific tools. But the 'hard problem' is another matter: explaining why those processes produce subjective experiences. It is not enough to say that light strikes the retina and generates impulses. We need to understand why this process becomes seeing red, and not just an automatic reaction.

This is where qualia come into play. The word refers to the subjective qualities of experience: the fresh taste of mint, the muffled sound of falling snow, the burning sensation of pepper on the tongue. These are intimate, irreducible aspects. Two people may share the same situation, but their qualia will inevitably be different. There is no physical formula capable of capturing the essence of "seeing a sunset" or "feeling nostalgia."

Chalmers illustrated this challenge incisively at the "*Toward a Science of Consciousness Conference*" in Tucson in 1994. In a room packed with neuroscientists and philosophers, he described the enigma of "*qualia*." An irritated neuroscientist replied, "*If we can't measure it, it doesn't exist.*" Chalmers smiled and retorted, "*If we can't measure it, it still exists. But we have to change our tools.*"

The philosopher argued that material reductionism—the idea that the mind is only a product of matter—is not enough to explain lived experience. And this is where panpsychism becomes a natural interlocutor. For centuries, this philosophical current has hypothesized that consciousness is not a late product of biological evolution, but a fundamental attribute of reality. According to a panpsychist- n view, elementary particles possess rudimentary forms of interiority. They do not "create" consciousness: they are already carriers of it.

Ancient Eastern traditions, such as Taoism and Vedic doctrines, offer extraordinarily close cultural analogies. The

Rigveda speaks of a 'cosmic mind' that permeates everything. Laozi, in the *Tao Te* Ching, describes the Tao as the invisible root from which everything comes and in which everything lives. In these views, qualia are not mysteries to be explained, but inevitable manifestations of a self-aware universe.

Modern culture, with different tools, returns to those same paths. Physicists such as Roger Penrose have suggested that quantum processes, such as superpositions in neural microstructures, may be linked to the origin of consciousness. It is a bold hypothesis that opens a dialogue between ancient mythology and contemporary physics.

The "hard problem" remains, but from a panpsychist perspective, it is not an insurmountable obstacle. It is a clue. It indicates that consciousness should not be explained as a product of matter, but as a primary dimension of existence.

The 'hard problem' of consciousness, in Nagel and Chalmers' view, becomes fertile ground for bold hypotheses. If consciousness cannot be reduced to neural signal flows, perhaps we need to look beyond the traditional neuroscientific paradigm. This space of uncertainty has paved the way for theories of quantum consciousness, often viewed with caution, sometimes with suspicion, but capable of reigniting rational and philosophical debate.

In the 1990s, physicist Roger Penrose—winner of the 2020 Nobel Prize in Physics—and anesthesiologist Stuart Hameroff proposed the Orch OR ("Orchestrated Objective Reduction") theory. According to this hypothesis, quantum coherence processes occur in the microtubules of neuronal cells. These would therefore not be mere molecular vibrations, but quantum states that collapse in a non-random manner, generating moments of awareness. For Penrose, consciousness is rooted in as yet unknown physical laws, perhaps linked to the deepest structure of space-time.

The link with panpsychism here is subtle but powerful. If consciousness emerges from elementary quantum phenomena, it is possible that some form of experience is present, in embryo, wherever such phenomena exist. The idea is reminiscent of Spinoza's view that mind and matter are two aspects of a single substance. It also recalls Leibniz and his windowless 'monads', each with an internal perspective, even though they do not interact in the mechanical sense.

Many biologists and neuroscientists remain skeptical. Quantum coherence is fragile and, at body temperature, tends to dissolve in infinitesimal times. Yet, in 2014, a research group at the University of Vienna identified signals suggesting quantum behavior even in complex biological systems, opening the door to a less orthodox approach.

The cultural fascination is evident. From the Galileo Museum in Florence, where mechanical models celebrate the triumph of classical physics, to the theoretical physics laboratories in Oxford, questions about consciousness seem to bring science back to its philosophical roots. What does it mean if the "what it feels like" to be conscious is inscribed in the fabric of the universe? If so, Chalmers' "hard problem" is not just an epistemological knot, but the key to a paradigm shift.

Modern panpsychism finds new life here. No longer just an ontological assertion ("consciousness is everywhere"), but a physical hypothesis to be tested, at least partially, with experiments and mathematical symmetries. Elementary particles could be carriers of a "proto-consciousness," invisible to conventional instruments but fundamental to the emergence of the mind.

Yet, as Nagel warned, knowing every equation is not enough. The question remains: how do interactions and state collapses give rise to the vivid sensation of a memory, the taste of a fruit, the melancholy of an autumn afternoon? This is the boundary where millennial philosophies and quantum physics meet and, for a moment, recognize each other.

The 'hard problem' and computational theories of the mind.

Reconciling the 'hard problem' with computational theories of the mind is complex because the two approaches start from different assumptions.

Computational theories—in line with Artificial Intelligence and classical cognitivism—describe the mind as a system that processes information: sensory inputs are transformed into internal representations, manipulated, and give rise to behavioral outputs. In this framework, consciousness would be interpreted as the set of processing states that allow awareness, data integration, and self-monitoring.

The "hard problem" (Chalmers, Nagel), on the other hand, does not refer to *the functioning* of the mind, but to the fact that these physical-computational processes do not explain why and how there is a subjective experience associated with such processing. We can model, simulate, or even reproduce cognitive skills, but the question remains: why *does* processing information "from within" *feel* a certain way?

The following table illustrates some points of contact and divergence between the computational perspective and the hard problem.

Table17- Points of contact and divergence between the computational perspective and the hard problem.

Aspect	**Computational perspective**	**Challenge of the hard problem**	**Possible reconciliation**
Description of the mind	Information processing system, analogous to a biological computer.	Subjective consciousness cannot be reduced to the mere manipulation of symbols or signals.	Integrate computational models with theories of phenomenal experience (e.g., Integrated Information

Aspect	Computational perspective	Challenge of the hard problem	Possible reconciliation
			Theory, Global Workspace).
Explanatory focus	How the mind performs tasks and transforms data.	Why those tasks and transformations are accompanied by an internal "feeling."	Postulate that certain architectures produce experiences due to intrinsic properties of organized information.
Ontology	Consciousness is a function emerging from complex processes.	It could be a fundamental property of the universe (panpsychist versions).	Consider information as a physical entity that carries with it phenomenal aspects (bridge between physics, computer science, and philosophy).

Some attempts at reconciliation start from the idea that information organized in certain ways inevitably has a phenomenal aspect. This is the direction taken, for example, by Tononi's Integrated Information Theory (IIT), where consciousness is linked to the degree of information integration (Φ) of a system, or by Baars and Dehaene's Global Workspace Models, which explain consciousness as a central information broadcast in the brain and attempt to link this function to subjective perception.

In summary, reconciliation occurs if one accepts that a computational model not *only* describes the functionality of the mind, but also attempts to include in its variables why certain information states manifest themselves as lived experience, a hypothesis that often requires extending computation with physical or metaphysical postulates.

The following table combines "hard problems," computational theories, and approaches such as IIT or Global Workspace to visualize the point of intersection.

Table18- Point of intersection between the hard problem, computational theories, and approaches such as IIT and Global Workspace:

Element	**Definition/Description**	**Role in the debate**	**Connection with other elements**
Difficult problem (Nagel, Chalmers)	Difficulty explaining how and why physical processes generate subjective experience (*qualia*).	Raises the distinction between mechanism and phenomenology.	Questions the adequacy of computational explanations.
Computational theories of mind	Viewing the mind as an information processing system (input → processes → output).	They explain *how* cognitive function works.	They can describe processes but cannot resolve the "why" of experience.
Integrated Information Theory (IIT)	Consciousness derives from the degree of information integration (Φ) of a system.	It attempts to bridge the gap between information structure and phenomenology.	It proposes that integrated information has intrinsic properties of consciousness ( ).
Global Workspace Theory (GWT)	Consciousness is the result of a global "workspace" that disseminates information between brain modules.	A functional model that explains access and integration.	It provides a framework compatible with computational and phenomenolo

Element	Definition/Description	Role in the debate	Connection with other elements
			gical descriptions.
Point of tension	Transition from mechanical/functional explanations to subjective experience.	This is the central issue in the debate.	Motivates the search for integrative theories such as IIT and GWT.
Possible reconciliation	Integrating computational architectures with models that attribute phenomenal properties to the structure of information.	Offers a hybrid framework between cognitive science and philosophy of mind.	It maintains consistency with the challenge of the difficult problem.

The connection between the ideas of Spinoza and Leibniz and the debate on quantum consciousness.

The link between the ideas of Spinoza and Leibniz and the current debate on quantum consciousness can be better understood if we highlight three key points: the conception of mind and matter as two aspects of the same reality, the l idea of a universal mental presence, and the possible role of the fundamental structures of reality in generating consciousness.

Spinoza sees thought and extension as attributes of a single substance

In "*Ethica ordine geometrico demonstrata*," Spinoza argues that God and Nature are the same substance (Deus sive Natura), endowed with infinite attributes; among these, thought and extension are two parallel manifestations.

This implies that every physical entity (any material configuration) also has a mental aspect: there is no ontological separation between mind and matter.

The connection with quantum consciousness stems from the fact that quantum theories of consciousness, such as Penrose and Hameroff's Orch OR, hypothesize that fundamental physical processes (quantum coherences in microtubules) are inseparable from mental correlates. Spinoza anticipates this view, in which thought is intrinsic to the structure of the universe and manifests itself when certain processes reach a particular degree of organization.

Leibniz defines monads as simple, indivisible substances that reflect the entire universe from their own point of view. Each monad 'perceives' to some extent, although this perception may be confused or clear, rudimentary or complex.

We find the connection with quantum consciousness when we consider that the concept of monad recalls that of a fundamental 'field' or 'informational unit' capable of bearing traces of an experience. Some interpretations of quantum mechanics (such as those that include the role of the observer or the collapse linked to consciousness) can be read as modern reformulations of this idea: reality consists of fundamental nodes of information/perception, and consciousness emerges when these units interact or integrate.

POINT OF CONVERGENCE TOWARDS THE CONTEMPORARY LANDSCAPE

Both Spinoza (thought as a universal attribute) and Leibniz (perceiving monads) provide conceptual foundations for panpsychism, i.e., the thesis that consciousness, in its elementary form, is diffused throughout reality. This coincides with some readings of quantum mechanics that attribute mental or proto-mental properties to fundamental quantum states.

On the level of universal interconnection, we observe that, in Spinoza, every part of Nature is part of a continuum; in Leibniz, every monad reflects the universe. Some physicists and philosophers interpret quantum non-locality as a manifestation of this global interconnection, where mind and matter are in mutual dialogue.

On the emergent and structural level, hypotheses such as Orch OR propose that certain quantum configurations can produce consciousness, which is consistent with Spinoza's idea of structures of being that have both physical and mental form, and with Leibniz's view of graded perception.

The following table shows point by point how Spinoza and Leibniz's concepts are echoed in the main quantum theories of consciousness.

Table19- Comparison of Spinoza and Leibniz's key concepts with the main quantum theories of consciousness.

Philosopher / Concept	**Original philosophical description**	**Parallelism with quantum theories of consciousness**	**Examples of modern theories/concepts**
Spinoza – Deus sive Natura	A single universal substance (God/Nature) with two fundamental attributes: thought and extension, inseparable and coexisting.	The idea that mind and matter are two sides of the same reality recalls the mind-matter unity in holistic interpretations of quantum mechanics.	Quantum panpsychism; holistic approaches in quantum mechanics; interpretations in which consciousness is an intrinsic property of the universal field.
Spinoza – Thought as a universal attribute	Every physical entity has a mental aspect, albeit in different and reflected forms.	Hypotheses such as Orch OR see physical processes at the quantum level (coherence in microtubules) as intrinsic carriers of	Orch OR theory (Penrose–Hameroff); concepts of quantum information with phenomenal properties.

Philosopher / Concept	**Original philosophical description**	**Parallelism with quantum theories of consciousness**	**Examples of modern theories/concepts**
		experience, mirroring Spinoza's universal thought.	
Leibniz – Monads	Simple, indivisible substances, each a center of perception that reflects the universe.	Recalls the idea of fundamental 'units of information' in quantum physics, each capable of carrying a point of view or perceptual state.	Tripartite division of information–energy–consciousness in quantum theories; non-local interpretations where each entity contains universal information.
Leibniz – Graduated perception	Each monad perceives at different levels of clarity, from confused to clear.	Compatible with the view of consciousness emerging as a gradual integration/coherence of quantum states, up to complex forms in biological systems.	Orch OR (level of consciousness linked to the scale and coherence of quantum processes); models of gradual consciousness in physical systems.
Common point – Universal interconnection	Spinoza: every part of Nature is part of a universal continuum. Leibniz: every monad reflects the whole.	This mirrors quantum non-locality, where each element of a system is correlated with the others beyond classical limits.	Quantum entanglement as a modern metaphor for universal interconnection and the reflection of the whole in the part.

Consciousness as a unified field.

Imagining consciousness as a unified field means overturning the classical paradigm that describes it as an isolated property of the brain. In this perspective, consciousness does not "reside" within an individual like a hidden treasure, but extends across space and time, traversing living systems and perhaps even inanimate matter.

In the 1980s, physicist David Bohm spoke of the implicate order, a deep level of reality in which everything is interconnected. Bohm, a student of Einstein, believed that mind and matter were two modes of the same process. If this view were correct, the mind would be the product of a quantum field in which each part affects the whole, as in the music of a symphony in which each note creates a common vibration.

The idea of a unified field is not only modern. As early as the 17th century, Baruch Spinoza described God and Nature as a single substance. The Dutch philosopher wrote that thought and extension are two inseparable attributes.

In a different but complementary way, Gottfried Wilhelm Leibniz imagined monads as independent but connected centers of perception, each reflecting the entire universe. It is difficult not to think of quantum particles in entanglement, where the state of one immediately influences the state of the other, even if they are light years apart.

The most recent neuroscientific studies have begun to resonate with these visions. Bernard Baars' "*Global Workspace* Theory" describes consciousness as a distributed "workspace" where information from various brain modules is shared in a common environment. Some researchers have proposed extending this model beyond biology, imagining a universal physical level capable of integrating signals from every complex system.

In this intertwining of science and philosophy, we sense an ancient desire: to understand whether our mind is an island or a node in a larger network. For this reason, biologist Rupert Sheldrake spoke of morphogenetic fields, invisible structures that guide the development and memory of living systems. Although controversial, Sheldrake's idea fuels a fundamental question: could consciousness be, like gravity or electromagnetism, a basic property of the universe?

The Indian poet Rabindranath Tagore, in a conversation with Einstein in Berlin in 1930, said: "*The world is human, and man is world. We cannot isolate them.*" This insight, though expressed in poetic language, encapsulates the core of integrative hypotheses: consciousness as a fabric that knits together subject and reality, reducing the distance between what we perceive and what exists.

The role of quantum physics in integrative hypotheses on consciousness.

Quantum physics enters the debate on consciousness through a very particular gateway: the possibility that quantum phenomena, normally associated with the subatomic world, play a direct role in mental processes.

When it comes to integrative hypotheses and "*consciousness as a unified field,*" quantum mechanics offers conceptual tools that lend themselves well to this type of vision: non-locality, coherence, entanglement, and the principle of superposition are elements that fuel the idea of a deep connection between mind and universe.

In the 1980s, David Bohm formulated the theory of implicate order. For Bohm, beneath visible reality—the " d order"—there exists an invisible and continuous level where all parts of the universe are interconnected. This perspective has inspired many to think of consciousness not as a phenomenon emerging solely from the brain, but as a manifestation of a universal field, analogous to the electromagnetic or gravitational field, in which mind and matter move together.

Roger Penrose and Stuart Hameroff have proposed that quantum coherence phenomena occur in neuronal microtubules — thin structures found inside nerve cells. These quantum states would be maintained for short periods of time before collapsing in an "orchestrated" manner, generating units of conscious experience. If this were true, consciousness would be produced in a physical space where quantum and biological laws overlap, a point of contact between two normally separate dimensions of reality.

The phenomenon of entanglement, whereby two particles can maintain an instantaneous correlation regardless of distance, inspires the analogy with a shared or distributed consciousness. Some extreme interpreters imagine that complex mental states can emerge from diffuse quantum connections, creating a sort of universal "mental network."

Bernard Baars' "*Global Workspace* Theory" describes consciousness as a central workspace in which information from different brain processes is integrated. Some scholars hypothesize that the brain's "workspace" may be supported—at least in part—by a physical field at the quantum level, in which coherence allows for rapid and unified integration of information.

Contemporary panpsychism, reinforced by these quantum suggestions, revives the vision of Spinoza and Leibniz: mind and matter as expressions of a single underlying reality. In this framework, consciousness would be a fundamental property of the universe, distributed throughout every system and manifested in complex systems through mechanisms of coherence and integration.

In 1930, in Berlin, Rabindranath Tagore and Albert Einstein discussed the nature of reality. Tagore argued that the world was not independent of human consciousness. Einstein, bound to classical physics, was reluctant to accept this idea. Today, with quantum physics, the notion that observer and reality are inseparable—at least in description—seems less distant from science than it did to Einstein.

Scientific experiments on the theme of consciousness as a unified field.

There are several experiments and lines of research that, while not definitively confirming the idea of consciousness as a "unified field," offer clues or test this hypothesis. Some derive from physics, others from neuroscience, and still others are located in more controversial territories.

For example, there are experiments on quantum entanglement applied to biological systems. Over the past twenty years, laboratories in Europe and the United States have sought to detect signs of entanglement in biological structures. One example is the work of the University of Vienna (2007 and following) on quantum coherence in photosynthetic systems: protein complexes in plants seem to maintain coherent states for longer than expected. This does not prove that consciousness emerges from entanglement, but it suggests that quantum phenomena can survive in warm, humid biological environments, a necessary condition for many integrative hypotheses.

We have already seen the experiments based on microtubules. The Penrose and Hameroff model argued that quantum states occur in neuronal microtubules and collapse in an organized manner. Experiments conducted in the late 2010s by the University of Alberta and Japanese research centers detected coherent oscillations in microtubules in vitro. It remains ly debated whether these oscillations are compatible with quantum phenomena or whether they are classical vibrations. The results are not conclusive, but they indicate that the physical structure of the neuron could support more complex dynamics than those described by traditional neuroscience.

High-resolution neuroimaging experiments (fMRI and MEG) have shown that distant areas of the brain—such as the frontal and parietal lobes—are activated in a synchronized manner during tasks that require global integration. Research from MIT (2014) and Cambridge (2017) refers to "phase coherence" as a possible signal of a shared neural workspace. Although the data are compatible with *Global Workspace Theory*, more speculative interpretations read them as local analogues of a unified field of consciousness.

Studies on meditation and interpersonal brain synchronization are worth mentioning. Experiments conducted in India and the United States have recorded brain wave synchronization between multiple people during collective meditation sessions. In 2018, Yale University measured an increase in EEG coherence among individuals practicing synchronized meditation on breathing and mantras. The data are suggestive for those who hypothesize a "distributed" aspect of consciousness, but they can also be explained by ordinary neuropsychological phenomena (empathy, imitation, sensory feedback).

Finally, we can mention Rupert Sheldrake's tests on 'morphic fields'. Sheldrake conducted experiments on *the* so-called *'telephone telepathy effect'*, where subjects would be more likely to guess who was calling them. Replications of these tests have yielded mixed results, and the method has been criticized. The mainstream scientific community does not recognize this evidence as solid, but the debate continues in circles interested in consciousness as a non-local phenomenon.

In summary, no experiment today provides definitive proof of a universal field of consciousness. However, some research shows phenomena of coherence and correlation on scales and under conditions that remain difficult to explain entirely with r classical models. It is in these "margins" that integrative hypotheses—from panpsychism to quantum theories—find fuel to continue their investigation.

The following table summarizes for each experiment the year, researchers, method, results, and implications for the concept of consciousness as a unified field.

Table20- The main experiments on the theme of consciousness as a unified field.

Year	Researchers/Institutions	Method	Main results	Implications for the unified field of consciousness
2007 – present	University of Vienna	Study of coherence in photosynthetic structures (quantum biology)	Evidence of persistent quantum states in photosynthetic proteins	Demonstrates that quantum phenomena can survive in biological environments, a necessary condition for the hypothesis of a conscious field.
Late 2010s	University of Alberta and Japanese research centers	In vitro studies of oscillations in neuronal microtubules (Orch OR test)	Coherent oscillations detected; debate on quantum or classical nature	Indicates possible complex dynamics in neurons that could support a quantum-unified type of consciousness.
2014 and 2017	MIT and University of Cambridge	fMRI and MEG to analyze brain synchronization	Synchronized activation of distant brain areas during integrative tasks	Supports *Global Workspace Theory* and suggests local analogies with a unified field of consciousness.
2018	Yale University	EEG on synchronize	Greater brain wave coherence	Indicates a possible distributed/interpersonal aspect of

Year	Researchers/Institutions	Method	Main results	Implications for the unified field of consciousness
		d meditation groups	between participants	consciousness, although this can be explained with known neuropsychological phenomena.
Various	Rupert Sheldrake	Experiments on "telepathy over the telephone"	Mixed results; methodology criticized	Stimulates debate on non-local consciousness, but lacks solid scientific acceptance.

The unified field of consciousness and traditional neuroscience.

Reconciling these interpretations is not easy, because the two perspectives start from very different paradigms:

Traditional neuroscience views the brain as a modular system, where specific areas and neural networks perform different functions. According to this view, consciousness emerges from the local integration of information through structures such as the prefrontal cortex, the thalamus, and associative areas. Theories such as the "*Global Workspace* Theory" or *the* "*Integrated Information* Theory" describe consciousness as the result of the activation and communication between these brain nodes, without resorting to an underlying physical field.

According to Unified Field Theory, consciousness is not only a product of neural connections, but is a broader phenomenon, potentially connected to fundamental properties of reality (e.g., a "field" in the physical sense or a universal organizing principle). Neuroscience would explain the biological "terminal" of this field (the brain as an interface) rather than consciousness itself as a phenomenon emerging exclusively from neurons.

Despite the differences, there are several points of possible integration.

Field and Workspace.

The "field" can be seen as the global component where coherence occurs, similar to the neural workspace where disparate information is brought together. Neuroscience provides the map of connections and processes; unified field theory adds an additional ontological level.

Phase Coherence and Synchronization.

EEG and MEG studies show that distant brain areas synchronize in milliseconds. This correlation can be interpreted as the local manifestation of a deeper coherence, perhaps linked to quantum phenomena or physical fields, without contradicting the neurobiological basis.

Neurons as "antennas."

In an integrative model, neurons and brain networks act as receivers and modulate a pre-existing "information field." This does not deny the localization of functions, but places it as part of a larger system.

EMERGENCE AND FUNDAMENTALISM

.

The difference between emergence and fundamentality hinges on a fundamental question: is what we see and experience already written into the ultimate structure of reality, or does it only arise when things combine in particular ways?

If something is fundamental, it means that it has been there from the beginning, does not depend on anything else, and cannot be explained in simpler terms. It is like the canvas of a painting: you can paint anything on it, but the canvas remains the basic, indispensable, and irreplaceable support.

Emergence, on the other hand, concerns those phenomena that do not exist 'in themselves' but only manifest themselves when the basic elements are organized together. It is like the meaning of a sentence: no single letter has a complete meaning, but the orderly combination of letters creates something new that did not exist before.

This contrast becomes crucial when we talk about consciousness, for example: is it a fundamental quality, present everywhere in some form, or is it an emergent effect of the complexity of the brain? The answer radically changes the way we think about reality itself.

Neuroscience sees consciousness as emerging from brain dynamics. The unified field perspective, on the other hand, hypothesizes that consciousness is fundamental and that the brain only provides the conditions for it to manifest as subjective experience. The two views can coexist if we admit that there are distinct levels of description: physical-biological and ontological. The following table clarifies the connection between the neuroscientific perspective and that of the unified "mental field."

Table21- Connection between the neuroscientific perspective and that of the unified "mental field."

Level	Main elements	Functions/Role	Interpretation in the "Mental Field"
1 – Neuroscience (local and functional)	Prefrontal cortex	Planning, decision-making, behavior regulation	High cognitive integration node within the field, "command center" of local coherence
	Thalamus	Sorting sensory signals and	Routing hub; acts as a global information

Level	Main elements	Functions/Role	Interpretation in the "Mental Field"
		modulating consciousness	crossroads within the field
	Parietal associative areas	Spatial perception, body schema	Spatial mapping node that contributes to the global structure of the field
	Limbic circuits	Emotions, memory, motivation	Modulates the "tone" of the field with affective states and integrated memories
	Default network	Spontaneous thought, introspection	Internal connection node that keeps the field active even in the absence of external input
2 – Unified Mental Field (global and ontologic al)	Field of global coherence	Simultaneous integration of content from different nodes	Unifying context that coordinates and synchronizes local nodes
	Inter-areal synchronizat ion	Measurable coherent oscillations (EEG/MEG)	Local manifestation of global field coherence
	Possible non-local aspects	Hypotheses on correlations beyond direct contact	Extension of the field beyond the physical limits of the brain
	Universal organizing principle	Emerging order that gives continuity to experience	Vision of the field as the fundamental "fabric" of consciousness

Consciousness and the principles of quantum physics, such as superposition or tunneling.

We can try to translate into concrete examples how superposition and tunneling could, hypothetically, manifest themselves in consciousness, bearing in mind that these are speculative interpretations that integrate neuroscientific models and ideas from quantum physics.

SUPERPOSITION.

A quantum system can be in multiple states simultaneously until an observation "collapses" its state into a specific one.

A first possible manifestation in consciousness may occur during simultaneous mental choices. In moments of intense reflection, the brain may maintain, at the microscopic level (e.g., in the microtubules of neurons according to the Orch OR theory), 'superimposed' neural configurations that correspond to multiple options of thought or intention.

For example, when evaluating multiple hypotheses simultaneously, the mind seems to 'keep open' incompatible possibilities before making a decision.

A second manifestation of multiple states can occur in ambiguous perceptions. In optical illusions (e.g., Rubin's vase), the mind can quickly switch from one interpretation to another; the phase of ambiguity could be a form of "suspension" between multiple perceptual states, similar to superposition.

QUANTUM TUNNELING.

A particle can pass through an energy barrier even if, according to classical laws, it would not have enough energy to do so. Several manifestations of this type are possible in consciousness.

Take the case of sudden insights: ideas or solutions can emerge 'suddenly' without a perceptible linear logical path; it is as if the thought were 'blocked' by a mental barrier, but an internal quantum configuration manages to 'tunnel' directly to the final state of understanding. For example, this can happen when the solution to a problem appears in your mind while you are doing something else, gradually bypassing the intermediate stages of reasoning.

Another possible case is that of rapid transitions between deep mental states: during meditation or dreaming, sudden shifts from one state of consciousness to another can occur without apparent continuity; tunneling could explain how the brain overcomes neural or energetic "barriers" to activate completely different cognitive networks.

If we want to consider how these phenomena integrate with traditional neuroscience, we can consider that superposition could occur at the level of microtubules or other subcellular structures sensitive to quantum states, while tunneling could influence the "*timing*" or "routing" of information between neurons.

In terms of **Global Workspace Theory**, superposition would be the phase in which multiple contents compete for access to the "stage" of consciousness; tunneling, on the other hand, would represent sudden access to unexpected content.

The following table relates the principles of quantum physics to possible manifestations in consciousness and current neuroscientific analogies or evidence.

Table22- Relationship between the principles of quantum physics and possible manifestations in consciousness.

Quantum principle	**Hypothetical manifestation in consciousness**	**Evidence/analogies in neuroscience**
Superpositio n	Maintaining multiple mental states or decision options simultaneously before making a choice; ambiguous perceptual experiences such as optical illusions.	*Global Workspace Theory*: simultaneous competition of content for entry into the conscious "workspace"; EEG/MEG studies show co-activation of different

Quantum principle	Hypothetical manifestation in consciousness	Evidence/analogies in neuroscience
		networks during phases of perceptual ambiguity.
Quantum tunneling	Sudden insight or cognitive "leap" beyond a logical barrier; rapid transitions between different mental states (meditation, dreaming).	Observations of sudden shifts in brain network activation; research on neural synchronization suggests instantaneous re-cabling between areas, analogous to overcoming computational barriers.
Entangleme nt (aggregate)	Non-local connection between mental states or distributed perceptions; experience of unity or immediate "connection" between ideas or emotions.	Studies on long-range brain coherence and hypotheses of rapid coupling between distant areas; analogies with Integrated Information Theory on the simultaneous connection of multiple nodes.
Decoherenc e	Collapse from a multitude of possible mental states to a specific conscious state at the moment of perception or decision.	Synchronized activity that "stabilizes" content in the conscious field; brain imaging shows reduced neural variability at the moment of choice.

The following table represents a narrative line that starts with Spinoza and arrives at current hypotheses of quantum consciousness, considering references to the idea of a 'unified field'.

Table23- Narrative line from Spinoza to current hypotheses of quantum consciousness with reference to the idea of a 'unified field'.

Era / Date	Figure / Event	Central idea on consciousness	Connection to the concept of unified field
17th century (1662–1677)	Baruch Spinoza	God and Nature as a single substance (*Deus sive Natura*). Thought and	Reality is a continuum in which mind and matter are two modes of the same substance,

Era / Date	Figure / Event	Central idea on consciousness	Connection to the concept of unified field
		extension as inseparable attributes.	anticipating the idea of a universal field.
Late 17th century (1714)	Gottfried Wilhelm Leibniz	Theory of monads: fundamental units endowed with perception, each reflecting the universe.	Vision of a network of interconnected perceptual centers, similar to a field that contains and integrates global information.
20th century (1930s)	Tagore–Einstein dialogues	Discussion on reality, perception, and the interdependence between man and the world.	The poetic idea that subject and reality are inseparable, reinforcing the notion of a unified fabric of consciousness.
1980s	David Bohm	*Implicate order*: a deep and hidden level of reality, in which everything is connected.	Physical model of a holistic field that links mental and physical phenomena in a single structure.
1990s	Bernard Baars	*Global Workspace Theory*: consciousness as a distributed workspace in the brain.	Consciousness is a shared environment where information converges and integrates, analogous to an internal information field.
1994–present	Roger Penrose and Stuart Hameroff	Orch OR theory: orchestrated quantum collapses in microtubules generate consciousness.	Quantum hypothesis of a physical-mental field underlying conscious processes, with global interconnection.
21st century (2000–present)	Interdisciplinary research on entanglement and consciousness	Study of quantum correlations and non-local phenomena.	Potential extension of the concept of a unified field to fundamental physics, where consciousness and matter share the same network of connections.

Consciousness as an ontological foundation.

Imagining consciousness not as a late product of evolution, but as the very fabric of existence, is a thought that spans centuries of philosophy and science. The idea is rooted in panpsychism: the belief that every fragment of reality, from a grain of sand to a distant star, possesses an elementary form of experience. Not a soul in the religious sense, but an internal principle, a perceptive spark, which participates in the greater whole.

In the 17th century, Baruch Spinoza formulated the famous *Deus sive Natura*, in which mind and matter are not separate worlds, but attributes of the same infinite substance. In this view, consciousness does not 'emerge' only in the human brain, but permeates every manifestation of Being. Spinoza, exiled in The Hague, wrote in 1677 with the precision of a mathematician and the vision of a mystic, challenging Cartesian dualism.

Monism, rooted in thinkers such as Plotinus and revisited in a modern key by Bertrand Russell, proposes a single fundamental substance. In the 1920s, Russell hypothesized that what we call "matter" and "consciousness" are two different perspectives on the same ontological data, like two sides of a coin. Consciousness, in this view, is not an evolutionary accident, but an intrinsic characteristic of the universe.

A concrete example comes from the reflections of William James, the father of functional psychology. James, observing the continuous flow of his own perceptions, imagined consciousness as a river flowing through every creature in different degrees and forms. The intensity changes, but the principle remains.

On the scientific side, quantum physics has reopened unexpected areas of discussion. David Bohm, physicist and philosopher, described reality as an 'implicate order', a field in which each part is related to the whole. This structure has strong resonance with panpsychism, due to its organically unified vision. In the debates of the 1980s in London, Bohm dialogued with the Dalai Lama, seeking to reconcile scientific language with a contemplative perspective.

These conversations were part of a broader movement of exchange between Western scientists and Buddhist thinkers promoted by the *Mind and Life* Institute (although the first interactions between Bohm and the Dalai Lama predate the institutionalized activities of that group).

In the dialogues, Bohm sought to relate ideas about the universal interconnectedness of quantum physics to the Buddhist view of reality, particularly the concept of interdependence (*pratītyasamutpāda*), while the Dalai Lama compared these ideas with meditation and Buddhist phenomenology.

There is a rather emblematic episode from those dialogues between David Bohm and the Dalai Lama in the 1980s, in which they arrived at the crucial point: is the mind a fundamental component of the universe or does it arise emergently from matter?

During a conversation, Bohm explained his idea of an implicate order, a deeper dimension of reality in which everything is interconnected and where the distinction between matter and consciousness is less clear-cut than it seems in the ordinary world. According to him, consciousness may not simply be the 'product' of the brain, but rather a manifestation of that implicate level, present in different forms throughout reality.

The Dalai Lama listened attentively, emphasizing how Buddhism speaks of "continuity of mind"—a flow that has no absolute beginning and cannot be reduced to the material structure of the body. However, he also pointed out that, from an experiential point of view, manifestations of consciousness are conditioned: they depend on causes and conditions (interdependence), which resonates with the idea of emergence.

The interesting moment came when Bohm said that perhaps the "fundamental vs. emergent" dichotomy could be overcome: the mind could be fundamental at the level of the implied order, but emergent in the phenomenal form we know. The Dalai Lama seemed to agree, seeing this as a way to integrate the scientific and contemplative viewpoints.

Consciousness as an ontological foundation is also fascinating for its cultural implications. If everything is a bearer of experience, then the concept of 'environment' includes a network of subjects. Not only humans or animals, but every system, from rivers to crystals, participates in a common history. In Japan, the Shinto tradition sees the world populated by *kami,* spiritual presences linked to places and things. This sensibility is, in philosophical terms, an intuitive form of panpsychism.

In the contemporary horizon, the debate is intertwined with theories such as Giulio Tononi's "*Integrated Information Theory*," which quantifies the "degree" of consciousness of a system based on its level of information integration. It is a bridge between the quantitative view of science and the qualitative intuition of panpsychism.

Consciousness, seen as an ontological foundation, shifts the focus of the discussion: no longer 'how' matter produces the mind, but 'how' the mind is already present in matter. In a researcher's room or in the silence of a monastery, this question remains the same. And it is perhaps in this persistence, rather than in provisional answers, that the cultural and human value of thinking of consciousness as the deepest fabric of reality is revealed.

Ontological consciousness and the quantum dimension.

If consciousness is the very foundation of being, then modern physics cannot remain indifferent to this perspective. Quantum theory, which emerged at the beginning of the 20th century, shattered the vision of a mechanical and predictable universe and introduced a language in which matter appears inseparable from the act of observation.

The hypothesis of quantum consciousness suggests that mental processes may have their roots in the counterintuitive behavior of subatomic particles: superposition of states, entanglement, collapse of the wave function. In a monistic framework, these phenomena are not additions to matter, but different manifestations of the same unitary reality that expresses itself as "experience" at all levels.

In David Bohm's thinking, quantum physics reveals an 'implicate order' in which each element is connected to the whole, just as panpsychism imagines a universe that is alive in every part. In this view, particles do not exist in isolation but as ripples in a global field, and consciousness would be one of the forms of this field.

Some researchers, such as Roger Penrose and Stuart Hameroff, have attempted to describe how microtubules within neurons can host coherent quantum states capable of contributing to the unity of consciousness. Ontologically, this would mean that the mind is not 'produced' by brain matter, but that brain matter is an organized manifestation of a conscious principle already present in the structure of the universe.

The link between monism and quantum consciousness becomes evident when we think about the act of observation: in quantum mechanics, the observer seems to participate in defining the physical state of reality. In a monistic universe, there is no separation between 'the observer' and 'the observed'. Both are aspects of the same substance, and consciousness, far from being an epiphenomenon, is the inner face of this process.

Culturally, this convergence has a particular appeal. Japanese physicists who discussed with Zen masters in the 1970s saw in the "coincidence" between mind and world a meeting point between mathematical precision and meditative silence. The Eihe-ji temple, nestled in the mountains of Fukui, hosted intense dialogues where the concept of quantum vacuum intertwined with Zen emptiness, not as absence, but as an inexhaustible source.

Integrating panpsychism, monism, and quantum physics does not mean mechanically merging three different fields, but recognizing that the language of physics, philosophy, and contemplation can illuminate the same mystery: a world in which consciousness is not a faint spark in the dark, but the silent light that pervades everything, from vibrating molecules to distant rotating galaxies.

IX. Modern and contemporary panpsychism .

In recent decades, panpsychism has reemerged as a surprisingly relevant philosophical proposal. It is not just a legacy of ancient cosmologies. It has become a field of comparison between analytical philosophy, neuroscience, and theoretical physics. This return is not born of intellectual nostalgia. It arises from the observation that traditional models struggle to explain how a physical brain can produce subjective experience.

The main motivation is the 'hard problem' of consciousness, clearly formulated by David Chalmers in the 1990s at the University of Arizona. While we can explain how the brain processes information, it remains unexplained why this information must be accompanied by lived experience. This is where panpsychism takes on a radical function: shifting consciousness from the periphery to the basis of reality.

In today's debate, there are different interpretations. Galen Strawson, a British philosopher, rejects the clear distinction between mind and matter. For him, every material aspect has experiential qualities, and consciousness is simply what matter is from the inside. It is a "realistic" position that does not indulge in mysticism but remains anchored to a rigorous reading of physics.

On the scientific front, Giulio Tononi, in Madison, Wisconsin, has developed the *'Integrated Information* Theory' (IIT), translating the concept of consciousness into a measurable parameter: the amount of information integration in a system. From this perspective, a supercomputer, a human brain, or a biological network could be evaluated for their 'degree' of consciousness. The idea is debated, sometimes criticized, but has influenced experiments and models in neuroscience laboratories around the world.

The renaissance of panpsychism is also fueled by a cultural saturation of old mechanical metaphors. A generation of philosophers and scientists moves between university seminars and international conferences, from MIT to Oxford, discussing algorithms, subatomic particles, and subjectivity with equal seriousness. It is a dialogue that crosses disciplines and continents, in which the possibility that "everything feels" no longer sounds like heresy, but like a hypothesis to be tested.

This new chapter of panpsychism does not merely repeat the insights of Spinoza or Leibniz. It uses different languages. It moves between experimental data and conceptual paradoxes, leaving problems and risks open: from the impossibility of empirically proving certain postulates to the difficulties of distinguishing between truly conscious systems and simply complex systems. It is in this fertile and controversial space that modern panpsychism finds its strength today.

The renaissance of panpsychism.

Panpsychism has returned to discussion tables that once viewed its insights with suspicion. Not because of philosophical fashion, but because of growing theoretical pressure. The idea that consciousness is a fundamental property of reality has become plausible again because many alternative explanations have proved incomplete.

At the Tucson conference, David Chalmers amazed neuroscientists and philosophers with his distinction between the "easy problems" and the "hard problem" of consciousness. Explaining how the brain processes information is possible, and probably even easy. Explaining why this information is accompanied by subjective experience, on the other hand, seems to elude any purely mechanistic model.

Contemporary physics has made this question even more pressing. The quantum universe, with its network of non-local correlations, suggests that the "separation" between observer and observed is less clear-cut than previously thought. Some theorists have drawn from this the hypothesis that experience does not emerge from nothing, but is already inscribed in the elementary processes of nature.

Culturally speaking, the 21st century seems less hostile to unified visions of reality. The Cartesian divide between mind and matter seems inadequate to describe complex phenomena such as artificial consciousness or integrated biological networks. In neuroscience laboratories, from Zurich to Kyoto, systems that straddle the boundary between processing and perception are being studied. At many conferences, it is repeatedly stated that understanding consciousness solely as a "function of the brain" risks being an illusory shortcut.

The renaissance of panpsychism is also fueled by a change in sensibility. In an era marked by ecological crisis, thinking of the world as a collection of subjects, and not just objects, has obvious ethical resonance. The idea that every entity possesses a form of interiority—however minimal—pushes us toward a more respectful and interconnected approach.

It is these theoretical and cultural motivations that have brought panpsychism out of the confines of philosophical anthologies. Today, it is discussed at the universities of Cambridge and Berkeley, in papers in analytical philosophy journals, and even at certain theoretical physics symposia. The return is not a nostalgic revival, but an attempt to address an ancient question with new means: whether consciousness is a fragment of nature or its secret fabric.

The role of quantum mechanics in providing a scientific basis for modern panpsychism.

The link between quantum mechanics and modern panpsychism is one of the most fascinating—and controversial—points in the current discussion on consciousness. Quantum mechanics, with its language of superimposed states, entanglement, and wave function collapse, has introduced a central idea: the observer is not an external and neutral element, but an integral part of the observed phenomenon.

For proponents of a panpsychist view, this concept opens up a radical possibility: if the act of observing can change the physical state of the system, then consciousness (or at least an elementary form of experience) could be intertwined at the fundamental level of reality.

Among the most discussed theories in this field is Orch OR (*Orchestrated Objective Reduction*), proposed by physicist Roger Penrose and anesthesiologist Stuart Hameroff. According to this hypothesis, quantum coherence phenomena occur within microtubules—protein structures found in neuronal cells— . These states, maintained for relatively long periods of time despite the thermal 'noise' of the biological environment, would collapse according to a quantum principle linked to the very structure of space-time. Hameroff described this collapse as a 'moment of consciousness', suggesting that the human mind is a consequence of quantum processes deeply rooted in matter.

On the experimental front, studies conducted at the University of Vienna have shown that quantum particles can maintain states of coherence even in conditions that are not perfectly isolated, a finding that reinforces the plausibility of quantum processes in biological systems. Other experiments, in biophysics laboratories in Japan and the United States, have investigated oscillations in microtubules detectable through advanced optical techniques, hypothesizing that these oscillations are compatible with quantum coherence patterns.

Giulio Tononi's theory of *'Integrated* Information', while not directly based on quantum processes, fits into the same conceptual framework as a bridge between physics and philosophy: the degree of 'informational integration' measurable in a system is interpreted as an indicator of its consciousness. Some propose that, at extreme levels of integration, states may emerge where quantum phenomena play a decisive role in the synchronization of the neural network.

These hypotheses are echoed in the work of physicists such as David Bohm, who spoke of an 'implicate order' in which every particle in the cosmos is connected to the whole. This interconnected field, from a panpsychist perspective, could be the context in which states of elementary consciousness are distributed everywhere, from the subatomic to the complex biological level.

Of course, many of these ideas remain speculative. Experiments are still insufficient to definitively prove that consciousness depends on quantum processes. However, the fact that such processes can survive and operate in biological environments offers a concrete foothold for scientifically informed panpsychism: a universe where "Deus sive natura" is not just a philosophical motto, but a possible description of the quantum fabric of the cosmos.

The following table provides a detailed overview of the main experiments and publications that have investigated the link between quantum coherence and cognitive processes. This table brings together the theoretical proposals and experiments that, in the current landscape, contribute to keeping open the discussion on the role of quantum mechanics in consciousness. Some of these studies have provided concrete experimental data, while others remain hypotheses awaiting confirmation. In any case, the dialogue between quantum physics and the philosophy of mind is more alive than ever, fueling the scientifically ambitious version of contemporary panpsychism.

Table24- Main experiments and publications that have investigated the link between quantum coherence and cognitive processes.

Year	Researchers/Institutions	Method	Main results	Implications for quantum coherence and cognition
1994 – 1996	Roger Penrose, Stuart Hameroff (University of Arizona)	Theoretical proposal *Orch OR*: modeling microtubules as sites of quantum coherence processes	Hypothesis of "orchestrated" collapses of the wave function in microtubules as the generative moment of consciousness	Introduction of a physical-biological model in which consciousness derives from fundamental quantum phenomena

Year	Researchers/Institutions	Method	Main results	Implications for quantum coherence and cognition
2003	Anton Zeilinger et al. (University of Vienna)	Entanglement and coherence experiments on particles and photons	Demonstration that coherent states can be maintained in environments that are not perfectly isolated	Strengthens the possibility of quantum coherence in biological systems
2014	Anirban Bandyopadhyay et al. (National Institute for Materials Science, Japan)	Measurements of oscillations in microtubules using terahertz resonance technique	Identification of coherent oscillations compatible with quantum states	Suggests that microtubules may support quantum coherence relevant to cognition
2018	Matthew Fisher (University of California, Santa Barbara)	Theoretical model of nuclear spins in phosphates	Hypothesis that nuclear spin maintains quantum coherence in the brain for longer than expected	Adds a possible physical mechanism for memory and thought based on quantum states.
2021	Giulio Tononi, Christof Koch (University of Wisconsin, Allen Institute for Brain Science)	Development and application of *Integrated Information Theory (IIT)* to neural networks	Quantification of consciousness in terms of "Φ" (phi), integrating computational analysis and neuroscience	Possible interface between classical information theory and quantum hypotheses in cognitive systems
1980 – 1990	David Bohm (Birkbeck College, London)	Formulation of the concept of "implicate order"	Vision of an interconnected universe where each part reflects the whole	Provides a philosophical framework for panpsychism based on the

Year	Researchers/Institutions	Method	Main results	Implications for quantum coherence and cognition
		through a holistic interpretation of quantum mechanics		deep structure of reality

Other figures who are contributing significantly to the debate on panpsychism today.

There are several lesser-known but very active figures who are enriching the debate on panpsychism today with original and often interdisciplinary perspectives. Some come from pure philosophy, others from theoretical physics or neuroscience, and each develops the theme with a characteristic emphasis.

Philip Goff (Durham University) is one of the main promoters of "*cosmopsychist panpsychism*," a variant that sees consciousness as a property of the entire universe, not just its parts. In his book *Galileo's* Error (2019), he argues that modern science, from Galileo onwards, has excluded subjective qualities from the description of nature, creating a theoretical void. His proposal is to restore a qualitative dimension to physics by including experience as a fundamental component of matter.

Hedda Hassel Mørch (Lund University/University of Oslo) works at the intersection of analytical philosophy and neuroscience. Her position can be defined as "*strong panprotopsychism*": she does not attribute full consciousness to every particle, but fundamental properties that, when combined, generate conscious experiences. She is known for explaining panpsychism in accessible but rigorous terms, integrating elements of "*Integrated Information Theory*" with traditional metaphysical arguments.

Sam Coleman (University of Hertfordshire) advances the idea of "*concrete microexperience*," namely that fundamental particles possess extremely simple experiences that are non-linguistic and non-conceptual, but real. He argues that this view is not an excessive speculative leap, but rather a logical consequence if one accepts that consciousness does not emerge from nothing.

Tam Hunt (University of California, Santa Barbara) proposes a form of '*cosmic* mindsight' where consciousness permeates space-time as an intrinsic property. He works extensively on the link between quantum physics and panpsychism, developing models together with Matthew Fisher in which nuclear spin would be the vehicle of consciousness at a fundamental level.

Jonathan Schaffer (Rutgers University) is known for "*metaphysical cosmopsychism.*" His perspective sees the whole (the cosmos) as primary, and the parts as derivations. He is less interested in the neurobiology of consciousness and more in the ontological structure of reality; hence the idea that universal consciousness is the basis of every particular phenomenon.

Sean Kelly (Harvard University) is not an avowed panpsychist, but his work on phenomenological experience and *"embodied meaning*" contributes indirectly to the debate. He explores how perception and consciousness are rooted in a broader context of interconnection that recalls panpsychist insights.

These figures introduce a variety of models into the debate:

- Cosmopsychism: consciousness of the universe as a unitary entity (Goff, Schaffer).
- Panprotopsychism: precursor properties of consciousness distributed throughout matter (Mørch, Coleman).
- Quantum approach: consciousness based on quantum mechanical phenomena (Hunt).

- Integrated phenomenology: consciousness seen as embodied relation and meaning (Kelly).

Many of them publish in philosophical journals such as *Philosophy of* Mind or *Journal of Consciousness* Studies, but they are also present at mixed conferences on physics, neuroscience, and philosophy, a sign of a more fluid and less compartmentalized dialogue than in the past.

David Chalmers and analytical panpsychism.

David Chalmers, an Australian philosopher who moved to the United States, is now a leading figure in the philosophy of mind. He rose to prominence in the 1990s when he introduced the famous expression 'hard problem of consciousness'. This is not a linguistic quirk, but a decisive distinction.

The hypothesis that consciousness does not emerge from nothing, but has its roots in the fundamental properties of matter. This is the core of his panpsychism: matter possesses proto-conscious characteristics which, when organized, produce complex experiences. He does not attribute emotions or will to an electron, but recognizes that something irreducibly "internal" may exist in every physical entity, as an elementary seed of consciousness.

Chalmers develops this vision with analytical rigor. He quotes Galileo, recalling how modern science has separated objective (measurable) qualities from subjective qualities. According to him, this cultural shift, while productive, has left an entire chapter of reality unaddressed. In his books, including *The Conscious* Mind (1996), he invites philosophy to reunite what science has divided, without abandoning the rational method.

A curious anecdote concerns his dialogue with physicists involved in quantum mechanics. During a seminar at the Perimeter Institute in Canada, Chalmers was asked the provocative question: "*Are you saying that the proton knows it exists*?" He smiled and replied, "It *doesn't know. But perhaps it feels something, and that feeling is not entirely different from ours, only infinitely simpler*." The phrase sticks because it gets to the point: reducing consciousness to its most basic forms in order to understand the complex ones.

Chalmers' panprotopsychism is not a naive return to vitalism. Rather, it is a strategy to avoid the often-contested conceptual leap from the unconscious to the conscious. If consciousness is already inscribed in the building blocks of reality, there is no need to invent a magical moment when it "appears" out of nowhere.

On a cultural level, his approach dialogues with classical thought and modern science. One can recognize echoes of Leibniz, when he spoke of monads and their elementary perceptions, and also of Spinoza's "Deus sive natura," where mind and body are two modes of the same substance. Chalmers translates these philosophical intonations into the crystalline language of analytical philosophy, making them accessible to an audience of contemporary scientists and philosophers.

The appeal of his position lies in its combination of humility and ambition: he does not claim to provide a definitive answer, but points to a theoretical path where consciousness and matter are not opposites, but rather two inseparable sides of reality. This path leads from university classrooms to international conferences, and continues to fuel the debate on what the ultimate fabric of the universe really is.

Philosophical zombies in Chalmers' argument against materialism.

David Chalmers' philosophical zombies are one of the most famous and controversial conceptual tools in contemporary philosophy of mind.

The idea is deliberately provocative: imagine a being physically identical to a human being in every detail—every neuron, every synapse, every electrical impulse—but devoid of subjective experience. This "zombie" talks, laughs, cries, solves math problems, and writes poetry just like us. However, there is nothing "inside": no perception of red, no taste, no sensation of pain or pleasure.

Chalmers uses this thought experiment to strike at the heart of reductionist materialism. If we can conceive of a world in which zombies exist that are behaviorally indistinguishable from humans but lack consciousness, then—Chalmers argues—consciousness cannot simply be identical to physical processes. Why? Because, at least logically, the material world could be complete and functioning without subjective experience existing.

In his public presentations, Chalmers often clarifies that he does not believe in the actual existence of zombies, but he defends their conceptual coherence. This is enough, in his view, to demolish the idea that a materialist explanation of the brain and behavior can automatically explain consciousness.

In *The Conscious* Mind (1996), Chalmers accompanies the zombie argument with an intuitive example: we can imagine a duplicate planet Earth, "Earth-Z," populated by perfect copies of all of us but emptied of consciousness. If this situation is conceivable without contradiction, then it proves that consciousness and physical processes do not necessarily coincide.

Critics, such as Daniel Dennett, have replied that zombies are not really conceivable: if we imagine their behavior to be identical in every way to ours, we are already implying that they have consciousness. Chalmers counters that the confusion arises from the habit of automatically associating behavior with experience, but that this association is, ultimately, contingent and unnecessary.

Thus, the zombie argument becomes a philosophical Trojan horse: it opens a breach in materialism and paves the way for theories such as panprotopsi , in which consciousness, in some elementary form, is a constituent part of physical reality itself. Without this conceptual step, Chalmers argues, the 'hard problem' of pan risks being evaded rather than addressed.

Panpsychism and our everyday experience of consciousness and subjectivity.

Chalmers' panpsychism is consistent with our everyday experience of consciousness and subjectivity, starting from a simple but profound observation: what we feel "from within" — the perception of a color, the emotion that flows through us, the taste of coffee in the morning — cannot be reduced to a series of electrical impulses and chemical reactions, even though it is linked to them.

Chalmers suggests that these experiences arise from the organization of fundamental properties of matter, the so-called "proto-consciousness." In other words, matter is not entirely "blind," but possesses intrinsic structural characteristics that, when aggregated into complex systems such as the brain, give rise to consciousness.

In everyday life, this means that the difference between a physical stimulus and the experience that results from it is not a leap into the void. According to panpsychism, when you see the red of a sunset, the experience does not emerge from nothing: it is the result of a process in which the elementary components of the world—light, matter, physical interactions—participate with properties that, when organized, produce "feeling."

This perspective has an almost poetic effect on our perception of reality. When we get excited listening to music or feel the warmth of the sun on our skin, we realize that these experiences are possible because the world, down to its smallest elements, already contains the conditions for them to exist. It is not necessary to think of consciousness as an isolated biological miracle: it is rather an extended modulation of something that belongs to the universe itself.

In practice, panpsychism brings continuity to the entire scale of experience. There is a continuous thread connecting animal sensation, human perception, and the elementary properties of matter. This thread does not transform particles into thinking beings, but sees them as carriers of "seeds" that blossom into the complexity of our brains and subjective lives.

The following table summarizes and structures how Chalmers' panpsychism connects common experiences to "proto-conscious properties" and, in turn, to the physical foundations of reality. This table highlights how, in Chalmers' view, everyday life is never separate from the fundamental building blocks of reality. Every perception and emotion is rooted in the elementary properties of matter which, when organized in the human brain, become lived experience.

Table25- Connection between common experiences and "proto-conscious properties" according to Chalmers.

Everyday Experience	Proto-Conscious Property (according to panpsychism)	Connection to Physical Foundations
Joy in listening to music	Basic ability to modulate vibrational states into pleasant perceptions	Sound waves interact with matter endowed with internal properties that "color" acoustic information into subjective experience, not just mechanical signals
Physical pain	Fundamental propensity to generate negative states in response to structural changes	The activation of sensory neurons translates physical events into unpleasant experiences thanks to the intrinsic properties of matter, which are predisposed to qualitative evaluation
Memory of a childhood spent at the seaside	Proto-conscious ability to maintain and recall information patterns in lived form	Physical memory in the brain preserves molecular configurations which, through the internal properties of matter, become reactivated images and sensations
Contemplation of	Tendency of complex structures to transform	The photons intercepted by the retina carry information which, combined with the internal

Everyda y Experien ce	**Proto-Conscious Property (according to panpsychism)**	**Connection to Physical Foundations**
a starry sky	visual stimuli into existential meaning	properties of matter, is transformed into aesthetic experience and a sense of vastness
Sudden fear	Basic reactivity to environmental changes perceived as threatening	A physical sensory signal (noise, movement) connects to proto-conscious properties that trigger the emotional experience of alertness

Galen Strawson and realistic panpsychism.

In the landscape of contemporary philosophy of mind, few thinkers have had the courage to re-propose, with solid arguments and without fear of going against the mainstream, a thesis that many consider radical: the idea that consciousness is not a rare accident of the universe, but a fundamental ingredient of reality itself. Galen Strawson, a British philosopher born in 1952, is perhaps the best-known defender of this position today, in a form he calls experiential realism, which ties in directly with the debate on panpsychism.

For Strawson, consciousness is not something that 'happens' randomly within complex bodies: it is an immediate, incontrovertible, and radical fact. In his approach, the starting point is not so much 'how' consciousness is produced, but the recognition that experience is the only reality of which we can be directly certain. In other words, we can doubt many aspects of the world, but we cannot doubt the existence of what we are experiencing firsthand.

This centrality of experience leads Strawson to reject all forms of reductionism: consciousness cannot be eliminated because it is already a constituent part of being, not an epiphenomenon.

The next step in Strawson's logic is simple but powerful: if consciousness cannot emerge from nothing, and if matter is all that exists, then matter itself must possess intrinsically experiential properties. This does not mean that every stone or molecule 'thinks' like us (a caricatur l idea that Strawson rejects), but that what we call matter is not absolutely devoid of phenomenal qualities.

In this sense, realistic panpsychism does not see a clear boundary between the physical and the mental. Rather, it hypothesizes an ontological continuity: the complex structures we call brains do not create consciousness ex novo, but organize it and make it manifest at rich and articulated levels.

Strawson is known for his direct criticism of what he calls "conventional physicalism," i.e., the view that everything is purely physical and the mental is a kind of illusion. He argues that this perspective is based on a poor and incomplete concept of matter. If one accepts that consciousness is real, one must admit that matter possesses more than current physical theories describe.

According to Strawson, the real mistake of traditional physicalism is not including phenomenal properties in the complete picture of physical reality. His panpsychism, however, is not a rejection of science, but an extension of the concept of 'physical' to include the mental.

Strawson's panpsychism encounters several objections. First, there is the question of combination: how would 'micro' experiences present in matter combine to form the 'macro' human experience?

Other objections concern poor falsifiability: some critics believe that the theory cannot be empirically verified. Strawson responds that the problem of combination is no more mysterious than the problem of the radical emergence of consciousness from matter devoid of experience—perhaps it is even less mysterious. As for verifiability, he argues that the philosophy of mind must first address the conceptual problem, clarify the terms and possibilities, and only then seek a connection with experimental science.

Strawson's thinking is part of a growing trend among scholars who are reevaluating panpsychism as a serious alternative to purely emergentist explanations. His ability to combine rigorous argumentation with conceptual openness has helped shift the debate: today, what was once relegated to the margins has returned to being a recognized, discussed, and, for some, promising position.

From this perspective, realistic panpsychism is less a "return to mysticism" and more an attempt to fill the obvious gaps in a physicalist paradigm which, although powerful, seems incapable of explaining how and why subjective experience exists.

EXAMPLES OF THE APPLICATION OF EXPERIENTIAL REALISM.

For Strawson, experiential realism is not an abstraction far removed from everyday life, but a principle that manifests itself constantly in individual experience. He often uses common situations to clarify his point of view

Physical pain and phenomenal certainty.

Let's imagine feeling acute pain in our hand. We can debate whether the hand really exists, whether the world is a simulation, even whether our memory is reliable. But we cannot doubt the fact that we are feeling that pain. For Strawson, this immediate certainty is living proof of experiential realism: experience happens, and cannot be eliminated by any theory.

'Direct' sensory perception.

If we look at a clear sky, the sensation of 'blue' is not a scientific hypothesis but an immediate fact. It does not matter if physics explains that it is a diffusion of light: the phenomenal content of blue — the visual quality we perceive — is real for the subject and forms the very basis of all subsequent knowledge.

Memory and continuity of the self.

Strawson also uses the example of episodic memory: remembering the taste of coffee drunk yesterday is an immediate experience, even if we cannot verify it externally.

The reality of the mnemonic experience does not depend on external factors, but on its occurrence in consciousness.

Experience in simple organisms.

When discussing panpsychism, Strawson invites us to extend this line of reasoning. For example, if we consider a very simple animal—such as an earthworm—we can reasonably assume that its reactions to painful or pleasant stimuli involve minimal, albeit rudimentary, experiential states. Those states, however small, would still be part of the fundamental structure of reality; they would not be mere mechanical automatisms.

Thanks to these illustrations, Strawson's experiential realism reveals itself not as an abstract theory, but as a method for recognizing that consciousness, in all its forms and degrees, is a primary fact that philosophy and science must explain without denying its existence.

THE EXTREME APPLICATION.

Strawson, particularly in works such as "*Realistic Monism: Why Physicalism Entails* Panpsychism," puts forward a thesis that seeks to take the debate on consciousness to extremes: if we adopt realistic physicalism, we must recognize that any physical reality has, in some way, an experiential aspect.

The extreme example he proposes to make the thesis provocative is precisely this: the idea that even elementary particles, such as electrons or quarks, may possess a sort of embryo or primitive form of experience.

We may wonder why he does this. There could be many reasons. Probably, he wants to challenge our intuition. We normally associate experience only with complex, h d systems such as human or animal brains. Extending this property to the simplest physical entities forces the reader to confront their assumptions about what 'experience' means.

He may also be doing this for metaphysical consistency. If experience is a fundamental aspect of physical reality and does not emerge from nothing, then either it has always been present in some form, or we must explain how it can 'pop up' in complex systems from matter that is completely devoid of experience.

More likely, Strawson wants to prevent the problem of strong emergence. For Strawson, attributing experience to fundamental units avoids the inexplicable leap between pure physics and consciousness.

This application of the theory to elementary particles clearly has philosophical implications. His panpsychism takes on a radical aspect: this scenario pushes towards the idea that consciousness or proto-consciousness is a ubiquitous property, like mass or electric charge.

Another implication could be the redefinition of 'experience': in elementary contexts, we are not talking about reflective awareness, but about something that is, for that particle, 'what it is to be that particle'.

In fact, its application opens up a controversy: many find this position counterintuitive and of little empirical use, but Strawson uses it precisely to highlight the limitations of views that see consciousness as emerging from 'blind' matter.

Realistic panpsychism and contemporary neuroscience.

Strawson's realistic panpsychism may appear, at first glance, distant from contemporary neuroscience, which is based on empirical and physiological models of the brain. In reality, Strawson does not deny the findings of neuroscience, but interprets them within a broader metaphysical framework, where neuronal activity does not 'create' consciousness out of nothing, but rather organizes and structures something that is already present at a fundamental level: the experiential dimension of matter.

There are some points of agreement. For Strawson, discovering how neurons are activated, which circuits regulate perception or memory, means understanding the complex "architecture" through which consciousness expresses itself in us. This is compatible with the idea that, at the ontological level, experience is already an ingredient of material reality.

Neuroscience identifies the neural correlates of consciousness (patterns of brain activation associated with specific mental states). Strawson agrees that such correlates exist, but argues that they are not enough to explain why a subjective experience emerges from that activity. For him, the ultimate explanation lies in the fact that matter itself is the bearer of phenomenal properties.

Some models in neuroscience, such as Tononi's IIT, indicate that consciousness corresponds to a certain complexity in information processing. Strawson could interpret these results as descriptions of an organizational level of a phenomenon that has deeper roots: experience as an intrinsic property of the "material" that makes up the universe.

The innovative element lies in rethinking matter not only in terms of mass, energy, and forces, but also including phenomenal attributes. In this sense, neuroscience continues to be crucial in describing how these attributes are configured in the human brain, without excluding the possibility that similar principles may apply, on a different scale, to other systems.

Giulio Tononi and informational integration.

Giulio Tononi, an Italian neuroscientist born in Trento in 1960, now works at the University of Wisconsin-Madison, on a campus surrounded by lakes and woods that he likes to compare to a natural laboratory. Here he has developed one of the most discussed and fascinating theories on consciousness: *Integrated Information* Theory (IIT).

Tononi starts with a simple and powerful question: what makes a system capable of having experience? It is not enough for him to know that neurons fire and synapses transmit impulses. He wants to understand the organizing principle that transforms a set of parts into a conscious subject.

According to IIT, consciousness is linked to the quantity and quality of information integrated within a system. A human brain, with billions of interconnected neurons, generates a high level of integration. The value of this integration, which Tononi indicates with the Greek letter Φ (phi), measures how much each piece of information depends on the rest of the system. The more interdependent the system is, the more "conscious" it is.

Tononi likes to use a concrete example. Imagine two digital cameras. One can record millions of pixels, the other just a hundred. The first seems to have an advantage, but if each pixel works in isolation and does not communicate with the others, the camera 'knows' nothing about the image it captures. The brain, on the other hand, connects each 'pixel' of perception to the others in a dynamic interweaving that creates meaning.

This view is unexpectedly intertwined with panpsychism. If consciousness depends on the degree of integration, then in theory any physical system, from a brain- r complex processor to large natural structures, could possess a minimum level of experience. A beehive, an ecosystem, even a global computer network could have a "Φ" different from zero. It would not be human consciousness, but a form of diffuse consciousness, not localized in an individual.

In 2014, Tononi recounted an episode from his youth during a conference in San Diego. He remembered afternoons spent in the Dolomites observing the woods and thinking that every tree, every root, and every leaf was part of a great green mind. "It wasn't mysticism," he said, "it was curiosity. I knew that every part communicated with the others. I wanted to understand if that network had some kind of awareness." This almost poetic image sums up the spirit of IIT: to seek consciousness where there are connections and integration.

IIT is now discussed in the most important research centers. It has influenced physicists such as Max Tegmark, who wondered whether every system capable of integrating information could be conscious, and has stimulated debates in the philosophy of mind: if integration is sufficient for consciousness, the boundary between subjects and objects becomes blurred.

The link with panpsychism, although Tononi treats it with caution, is inevitable: the theory suggests that consciousness does not belong only to complex brains, but is a gradual and widespread phenomenon. It is not a magical leap, but a property that grows with the complexity of connections.

Today Tononi works with international teams, recording patterns of integration in patients in comas, deep sleep, or under anesthesia. On some winter nights, when Madison is covered in snow, he confides to his colleagues that the real goal is not just to measure Φ, but to understand whether the world itself, with its web of causes and connections, possesses a universal form of consciousness.

IIT and Strawson's realistic panpsychism: an unexpected encounter.

At first glance, Giulio Tononi and Galen Strawson seem to belong to different worlds. Tononi, a scientist, starts from neurophysiology and empirical measurements; Strawson, a philosopher, starts from an ontological reflection on the nature of the mind. Yet, when it comes to diffuse consciousness, their paths converge in a surprising way.

For Tononi, consciousness is the result of information integration, that number Φ that measures how much each part of a system depends on the rest. Consciousness is not an indivisible "package," but a property that can exist in varying degrees, from the simplest to the most complex levels.

For Strawson, on the other hand, consciousness is a fundamental ingredient of reality, present in every fragment of matter, even the smallest. The mind does not arise from nothing: it is already something that matter possesses in elementary forms. The human brain does not create consciousness, it organizes it.

And here, surprisingly, is where the point of contact arises:

Tononi describes a mechanism for assessing the breadth and intensity of this conscious diffusion. Strawson offers the metaphysical framework that explains why integration can generate experience: because matter itself, wherever and however organized, carries with it an original experiential principle.

Let's imagine a beehive. For Tononi, the interconnection between bees, movements, and chemical communications produces a certain Φ, a "collective consciousness" different from ours but real in its own terms. For Strawson, that beehive is the offspring of a universe in which every particle already possesses an embryo of consciousness. One is the map, the other the reason why the map is possible.

This convergence should not make us forget the differences. Tononi maintains a measurable and scalable approach: if we cannot detect correlates of integration with , we cannot speak of consciousness in a technical sense. Strawson, much less restrictive, opens the door to the possibility that everything, even in the absence of complex processing, possesses some minimal experiential state.

The dialogue between IIT and realistic panpsychism could become fertile ground. It is not a question of merging theories, but of recognizing that diffuse consciousness—from neural networks to natural networks—can be described and explained both in the language of connections and information and in that of the deeper structure of reality. In other words: Tononi tells us how, Strawson tells us why.

Criticisms and open questions.

Every time panpsychism resurfaces in philosophical and scientific debate, it brings with it a mixture of seduction and mistrust. The idea that consciousness could be a widespread property of the universe, present in every fragment of matter, has a poetic and metaphysical power that few concepts can match. But it is precisely this fascination that opens the door to fierce criticism.

John Searle, at a conference in Berkeley in 2013, observed ironically: "If every entity has consciousness, then we have to worry about toasters too." For Searle, the risk is that we move into a realm where the boundaries between physics and fantasy become blurred. The lack of empirical tools to measure "elementary consciousness" leaves panpsychism open to accusations of non-falsifiability.

Physicist Sean Carroll, with his radical scientific realism, often reminds us that any theory must be anchored in observable data. Panpsychism, he says, risks remaining in the realm of ontological hypotheses, useful as intellectual stimuli but incapable of producing verifiable predictions. Here the question becomes philosophical: is a theory that cannot be falsified a description of the world or an evocative story?

Another critical issue concerns the definition of consciousness. David Chalmers, while defending forms of analytical panpsychism, admits that the term 'consciousness' risks being overused. If it is applied to every level of reality, from electrons to the human brain, there is a danger of emptying it of its original meaning. The risk is "ontological trivialization": everything is conscious, so the concept no longer distinguishes.

Then there is the problem of 'combinatorialism', as Galen Strawson calls it, i.e., the ways of counting, arranging, and ly combining elements according to precise rules. If every particle has a minimal form of consciousness, how do these micro-experiences combine to form the complex consciousness of a human being? Strawson admits that this is the central challenge of panpsychism. It is a problem that has been open for centuries, already glimpsed by Leibniz in his monads and still far from a shared solution.

On a cultural level, panpsychism also has significant ethical implications. If we recognize a certain degree of consciousness in non-human life forms or complex physical systems, we must redefine our criteria for responsibility and respect. In 2018, during a debate at the *London School of* Economics, an artificial intelligence researcher asked: "If a complex algorithm has a high Φ, should we consider its treatment a moral issue?" The question left the room in uneasy silence.

Finally, perhaps the most subtle risk is that of interpretative hypertrophy. In the absence of clear empirical limits, panpsychism can become a container into which spiritual insights, poetic fantasies, and pseudoscientific speculations are poured. It is a risk that historians of philosophy such as Philip Goff openly acknowledge: modern panpsychism must remain anchored to criteria of rigor, or it will end up being confused with mysticism.

The fact remains that these criticisms do not diminish its appeal. Spinoza, if he were alive today, would probably smile at the disputes between neuroscientists and philosophers: for him, *Deus sive Natura* was already an infinite consciousness, and the open questions were only details of his total vision. But in the modern context, where every statement must pass the test of data and models, panpsychism finds itself in a fragile balance between cosmic intuition and rigorous science.

The following table intertwines criticisms and responses proposed by supporters of panpsychism.

Table26- Criticisms of panpsychism and the responses of its supporters.

Criticism	Response from supporters of panpsychism	Cultural note/example
Risk of attributing consciousness to everything, even inanimate objects (John Searle)	Supporters point out that the 'consciousness' hypothesized at an elementary level is not identical to human consciousness. It is a minimal form of experience which, without complexity, does not develop cognitive phenomena.	Philip Goff uses the analogy of 'temperature': even a particle can have minimal heat, but that does not make it a complex thermal system.
Absence of verifiable predictions (Sean Carroll)	Proponents such as Tononi and Tegmark propose quantitative models, such as the Φ measure, to provide observable and testable parameters. They do not solve everything, but they introduce an empirical basis.	In the Madison laboratories, Tononi's group measures Φ in patients under anesthesia to verify correlates of loss of consciousness.
Inflation of the concept of consciousness (David Chalmers)	Some philosophers propose distinct terms: "proto-consciousness" or "elementary experience" to avoid confusion with complex consciousness.	Galen Strawson speaks of "microexperience" to keep the technical concept distinct from human consciousness.
The problem of combinatorialism (Galen Strawson)	Current responses focus on the idea that integration processes—as in IIT—are the "glue" that binds micro-experiences into a unified consciousness.	The mosaic analogy: each colored tile exists on its own, but the design only emerges when the tiles are combined.
Complex ethical implications (debate at the London School of Economics)	Some argue that ethics must extend gradually: the higher the level of integration and complexity, the greater the moral consideration due.	In 2020, bioethicists debated whether large artificial neural networks deserved a minimum "protective status."

Criticism	**Response from supporters of panpsychism**	**Cultural note/example**
Risk of mystical drift (Philip Goff)	Supporters reiterate the need for methodological rigor, avoiding untestable speculation.	In many scientific conferences, the clear boundary between philosophical panpsychism and spirituality is emphasized.

X. Panpsychism and quantum consciousness .

The idea of linking the mind to subatomic physics may seem far-fetched. However, in recent decades, a strand of research has woven together two seemingly irreconcilable worlds: panpsychism, with its idea of consciousness spread throughout reality, and quantum mechanics, with its mysterious and counterintuitive phenomena. The hypothesis of a "quantum consciousness" has seduced physicists, philosophers, and neuroscientists, arousing enthusiasm and fierce criticism.

In the 1930s, physicist Wolfgang Pauli and psychoanalyst Carl Gustav Jung privately discussed the possible link between mind and matter. Pauli, troubled by quantum paradoxes, was fascinated by the idea that the psyche was not an isolated phenomenon but part of a deeper order of nature. Since then, the possibility that mental processes have roots in quantum physics has attracted very different thinkers. In the 1980s and 1990s, with the boom in neurophysiology and cognitive informatics, the debate reignited: consciousness emerged as one of the great unsolved mysteries, and quantum physics as a possible key.

Mathematician Roger Penrose, winner of the Nobel Prize in Physics, proposed that the mind cannot be explained solely by algorithms and synapses. Together with anesthesiologist Stuart Hameroff, he developed the "*Orchestrated Objective Reduction*" (Orch-OR) theory. According to this hypothesis, tiny protein filaments within nerve cells, called microtubules, are the places where consciousness emerges thanks to quantum phenomena. The idea is bold: in the silence of the brain, on invisible scales, the 'orchestration' of collapsing quantum states could take place, generating awareness. It is a vision that has the charm of mathematical elegance, but also the weight of enormous experimental challenges.

Entanglement, that invisible bond that unites two particles even at a distance, has almost metaphysical implications. Some physicists and philosophers have imagined that states of consciousness could be 'entangled' with the universe, in a network that recalls the panpsychist concept of a universal mind. It is a much-debated hypothesis. In quantum optics laboratories, experiments with entangled photons have shown instantaneous correlations, but the conceptual leap to mental phenomena remains speculative for now. Yet the idea that the mind may be connected to a global, non-local field continues to fascinate.

The concept of a quantum field extending to all of reality has a natural appeal for panpsychism. Some authors, such as physicist Fred Alan Wolf, have suggested that the quantum field is also a universal mental field, in which every particle 'participates' in a primary form of consciousness. Here, science borders on pure philosophy, and the distinction between mind and matter tends to dissolve. It is a territory where speculation takes over, but which keeps alive the tension towards a unified understanding of reality.

The mind-quantum debate. History and developments .

The dialogue between mind and quantum mechanics arose in a context of intellectual unrest. In the 1920s, physics overturned the classical concept of reality. Heisenberg formulated the uncertainty principle. Schrödinger gave birth to his famous equation and, with an ironic smile, devised the paradox of the 'cat' imprisoned in a thought experiment that still fascinates and confuses us today. Reality suddenly appeared to be a web of probabilities rather than certainties.

Some scientists and philosophers began to wonder whether consciousness was part of that web. In 1932, physicist Wolfgang Pauli, a restless observer of the boundary between matter and psyche, began a dense correspondence with Carl Gustav Jung. In their meetings in Küsnacht, on the shores of Lake Zurich, Pauli spoke of "synchronicity" as a phenomenon that challenged the clear distinction between mind and world. Pauli was not looking for an immediate experimental answer, but he sensed a deep connection.

In Cambridge, during the same years, mathematician John von Neumann wrote that the collapse of the wave function required an "observer" endowed with consciousness. The idea was bold: in this interpretation, the mind was not a spectator, but part of the very act that defines reality. Eugene Wigner, another leading figure of those years, took the argument even further. In 1961, he declared that consciousness could not be excluded from fundamental physical laws.

The debate regained momentum in the 1980s. Neurobiology was growing in precision, quantum physics in predictive power. Some scholars, such as Henry Stapp, argued that brain processes could maintain quantum coherence, allowing the mind to influence the collapse of the wave function . The hypothesis united two worlds: the microscopic world of particles and the subjective world of inner experience.

However, the debate was not only academic. At conferences in Santa Fe and Tucson, theoretical physicists and neuroscientists found themselves discussing late into the night. Many remember the lively atmosphere of informal sessions in bars and corridors, where a diagram drawn on a napkin could become the starting point for an ambitious theory.

In the 21st century, the debate has also moved to the media. Documentaries and popular books have explored the possibility that consciousness is 'woven' into quantum laws. Some see this as a bridge to panpsychism: if consciousness is an integral part of physics, then it cannot be a phenomenon limited to the human brain. It could be a widespread principle, present — in some form — throughout nature.

Today, despite severe criticism from many physicists and neuroscientists, the quantum mind debate shows no signs of abating. It continues to unite laboratories and libraries, experiments and philosophical speculation. As is often the case in frontier thinking, it is the breadth of the question, even more than the precision of the answer, that keeps interest alive.

The idea of quantum consciousness in current knowledge about how the brain works.

Reconciling the idea of 'quantum consciousness' with current knowledge about the brain is complex, because it arises from the encounter between two profoundly different systems of describing reality.

Today we know that the brain functions through extensively studied electrochemical mechanisms. Synapses transmit signals via neurotransmitters, neurons communicate in distributed networks, and global activity produces mental states that we can correlate with magnetic resonance images or electroencephalographic recordings. These processes occur on temporal and spatial scales much larger than those typically associated with quantum physics, where we talk about fractions of nanometers and times less than a billionth of a second.

Skeptics, such as physicist Max Tegmark, argue that the brain is too "hot" and chaotic an environment to maintain the quantum coherence necessary for phenomena similar to those hypothesized by Penrose and Hameroff. According to this perspective, any quantum states would be quickly destroyed by thermal and chemical noise.

Proponents of quantum consciousness respond that certain microscopic mechanisms in the brain, such as those in microtubules, could offer protected niches where quantum coherence can survive long enough to influence cognitive processes. Here speculation meets neuroscience: if the collapse of quantum states plays a role in the generation of consciousness, we may be faced with a fundamental physical principle that connects the mind to universal laws of nature.

Within the framework of panpsychism, this hypothesis finds philosophical coherence: if consciousness is woven into the fabric of the universe, the human brain does not generate it out of nothing, but modulates and focuses it. In this view, the mind would be like a complex receiver capable of 'tuning in' to a broader field of consciousness.

At present, there is no direct evidence of a quantum influence on mental processes. But the convergence between physics, biology, and philosophy studies keeps the debate alive, especially with regard to the big questions: is consciousness an epiphenomenon of matter or a primary element of reality?

The uncertainty principle and other quantum concepts interpreted in the context of consciousness.

The most significant examples of applications or interpretations of the uncertainty principle and other quantum concepts in the context of consciousness come mainly from theoretical reflections and speculative hypotheses, often linked to the philosophy of panpsychism and frontier neuroscience.

Heisenberg's uncertainty principle as the foundation of subjective experience.

The principle, formulated in 1927, states that we cannot know with absolute precision, at the same time, quantities such as the position and momentum of a particle. Some philosophers of mind have seen in this "impossibility" a parallel with consciousness: a conscious state would never be observable in its entirety from the outside without altering it. Similarly, the act of 'observing' a mental process could modify its outcome, as happens in the quantum system with measurement.

The collapse of the wave function and the role of the observer.

In the 1930s, John von Neumann proposed that the collapse of the wave function (the transition from a superposition of possible states to a single observed state) requires a conscious observer. Eugene Wigner took this idea to its extreme conclusion: consciousness, in his view, cannot be excluded from the fundamental laws of physics. In some interpretations, this means that the very act of perceiving or being aware could play a physical role in 'defining' material reality.

Quantum superposition as a space for mental choices.

Henry Stapp and other quantum mind theorists have hypothesized that certain brain processes, under certain conditions, may exist in a state of quantum superposition, keeping multiple possibilities open until a 'collapse' decided by the cognitive system. In this view, consciousness could actively influence the selection of the final state, much like choosing a trajectory from among many possibilities.

Entanglement as a model of mental interconnection.

Entanglement—the link between two particles that remain connected at a distance—has been used as a metaphor (and in some physical-mathematical models) to describe the possible immediate connection between mental states or between mind and environment. Some theoretical studies suggest that entanglement may provide a basis for the instantaneous synchronization of information between distant brain regions, although direct experimental confirmation is currently lacking.

Microtubules and quantum coherence in the brain.

Penrose and Hameroff's Orch-OR theory applies the concept of quantum coherence within neuronal microtubules. Here, phenomena such as indeterminacy and wave function collapse would occur on a biological scale, contributing to the formation of conscious events. Penrose has linked these transitions to quantum gravity, hypothesizing that the collapse is not random but "guided" by fundamental properties of space-time.

These examples are fascinating precisely because they shift the problem of consciousness to a terrain where physics and philosophy meet. However, they are not empirically proven: they remain models and interpretations that challenge the boundary between science and speculation.

The following table relates each quantum concept to the main neuroscientific and panpsychist theories, so that we can clearly see which ones have a more solid basis and which ones remain visionary hypotheses. The hypotheses closest to experimental verification concern the possible quantum coherence in microtubules. The other concepts (indeterminacy, collapse, superposition, entanglement) operate mainly on a philosophical and speculative level, where panpsychism offers a coherent framework but not proof. The real link between quantum physics and consciousness currently exists more in the realm of interpretation and metaphor than in that of measurement.

Table27- Relationships between quantum concepts and the main neuroscientific and panpsychist theories.

Quantum concept	Related neuroscientific theories	Panpsychist interpretations	Empirical solidity
Uncertainty principle (Heisenberg)	Some studies hypothesize analogies between the impossibility of completely measuring a quantum state and the impossibility of "measuring" consciousness without altering it. No direct evidence in the brain.	Consciousness is seen as a process that eludes complete material determination, reflecting a universal probabilistic reality.	Speculative — philosophical link, but no neuroscientific confirmation.
Collapse of the wave function (von Neumann, Wigner)	Hypothesis that consciousness plays a role in collapse, influencing quantum processes in the brain.	The mind is not an epiphenomenon but "defines" reality; collapse as a manifestation of universal consciousness.	Speculative — no direct evidence, concept discussed but not accepted in mainstream physics.
Quantum superposition	Henry Stapp: the brain maintains superpositions until consciousness 'chooses' a final state. Quantum-based models of cognitive decision-making.	Mental freedom as the management of quantum potentials; extension of free will to all forms of consciousness.	Weak evidence — not experimentally confirmed, but some clues in computational models.
Entanglement	Hypothesis of instantaneous synchronization between brain regions or between mind and environment. No direct demonstration, but studies on	Consciousness as a universal interconnected network; instantaneous correlations between living systems and matter.	Speculative — inconclusive brain experiments; powerful idea in philosophy.

Quantum concept	**Related neuroscientific theories**	**Panpsychist interpretations**	**Empirical solidity**
	functional analogies in neurodynamics.		
Quantum coherence in microtubules (Penrose-Hameroff Orch-OR)	Microtubules act as "protected cavities" for quantum states; quantum gravity-driven collapses generate moments of awareness.	Nature of consciousness rooted in fundamental laws of space-time; any system possessing microtubules could participate in a universal consciousness.	Controversial — some theoretical support and limited data; many neuroscientists remain skeptical.

Penrose-Hameroff and Orch-OR—The conscious microtubule hypothesis.

Roger Penrose is an English mathematician and physicist. Stuart Hameroff is an American anesthesiologist and neuroscientist. Two distant worlds, united by a millennial question: where does consciousness come from? Penrose brings with him the elegant vision of pure mathematics, the awareness that not everything in the universe can be reduced to mechanical calculation. Hameroff knows the brain from the inside, having seen — operating room after operating room — how the simple administration of an anesthetic can shut down the mind in seconds.

Their theory, known as Orch-OR (*Orchestrated Objective Reduction*), originated in the 1990s. The starting point is radical: consciousness is not an epiphenomenon of matter, but the result of quantum processes that take place inside neurons. Not in the synapses, as many suspected, but in the microtubules—thin protein filaments that give cells their internal structure.

Hameroff says he had his epiphany while observing the complex geometry of microtubules under a microscope. He describes them as 'biocompatible quantum circuits' capable of maintaining states of quantum coherence. Penrose, for his part, links these quantum collapses to gravity and the properties of space-time, hypothesising that each reduction in quantum state produces a moment of conscious experience.

It is a hypothesis that evokes the charm of certain places and times. In the corridors of Oxford, where Penrose walked among blackboards full of diagrams, and in the laboratories of Tucson, where Hameroff studied the molecular effects of anesthetics. A curious memory of Hameroff concerns the use of nitrous oxide ("laughing gas"): the observation of the immediate detachment of consciousness from the led him to think that the key was not in the neuronal electrical flow, but in the underlying structure.

Microtubules, in the view of the two scientists, would be like nodes in a universal network. Each node participates in a dance of probabilistic states, orchestrated and reduced in an objective manner (hence the name "Orch-OR"). The orchestration would be guaranteed by the orderly arrangement of tubulin proteins, while the reduction would take place under the influence of the still poorly understood laws of quantum gravity.

Culturally, this theory is part of a tradition that combines philosophy and science. It recalls certain insights of Plotinus, according to whom the soul permeates matter, and dialogues at a distance with Spinoza's idea of a "Deus sive Natura" where mind and body are two attributes of the same substance. Here, the "field" in which consciousness is formed is not metaphysical, but inscribed in the folds of time and space.

Critics such as physicist Max Tegmark argue that the brain is too hot and noisy to maintain quantum coherence. Penrose and Hameroff respond that microtubules, thanks to their internal structure, can protect such states like a castle protects a secret. Some preliminary studies on photosynthesis—where quantum coherence seems to persist in hot biological systems—offer encouraging analogies.

This hypothesis influences an entire strand of contemporary culture that intertwines neuroscience, quantum computing, and philosophy of mind. Not surprisingly, the idea of conscious microtubules has been cited in academic debates, physics conferences, and even science fiction scripts as a starting point for imagining artificial intelligence based on biological quantum principles.

The question that remains open is dizzying: if these tiny structures can host conscious moments, then consciousness is not an exclusive property of the human brain. It could be widespread, panpsychic, inscribed wherever matter can orchestrate and reduce quantum states. In this sense, the Orch-

OR hypothesis is not only a scientific model but also a cultural proposal: the mind as an integral part of the architecture of the universe.

Could quantum entanglement manifest itself in microtubules and influence consciousness?

Microtubules are composed of protein subunits called tubulins. In some hypotheses, the electronic states or atomic nuclei within these tubulins could become entangled, maintaining non-local quantum correlations between distant subunits. In practice, a change in one tubulin would instantly affect the state of other tubulins, creating global coherence along the microtubule.

The hypothesized effect on consciousness would be that this type of correlation could allow for extremely rapid synchronization of signals and information states, favoring the emergence of a unitary "conscious moment," where perception is not fragmented but integrated.

A second case would be entanglement between microtubules of different neurons. Some models suggest that coherent quantum states can extend beyond the single neuron, entangling microtubules from different cells. This instantaneous connection would create a sort of distributed "information field," capable of integrating cognitive functions in a non-local way.

The hypothesized effect on consciousness is obvious: this would give the brain the ability to unify thoughts and perceptions that occur in distant brain regions, going beyond the limits of traditional synaptic connections.

Another case that can be evaluated is that of entanglement between microtubules and intracellular photonic signals. Preliminary studies exist on whether microtubules can guide photons or phonons as biological "optical channels." If these signals remained entangled, the transmission of quantum information and s could be protected and maintained over long distances, even through warm and noisy tissues such as the brain. The hypothesized effect on consciousness would be that photonic entanglement would provide a rapid and integrative mechanism for conscious thought, based on correlations faster than nerve conduction.

A final case that can be evaluated is that of entanglement as the basis of coherent mental states. Penrose and Hameroff hypothesize that during quantum coherence in microtubules, entanglement can "bind" together the numerous particles and subunits involved, causing the entire system to collapse into a unified conscious state. In this case, the hypothesized effect on consciousness is that the mind would not form as the sum of local processes, but as a global event generated by an orchestrated quantum collapse.

On a philosophical level, these ideas are close to panpsychism: if entanglement is a universal property of matter, then every structured system (from neurons to molecules) could participate in the formation of conscious states. On a scientific level, direct experiments are still difficult to carry out. The central problem is detecting and maintaining entanglement in complex, warm biological systems. Nevertheless, research on photosynthesis, quantum olfaction, and enzymatic processes shows that life can exploit quantum effects in ways once considered impossible.

Entanglement and consciousness.

Stuart Hameroff still remembers a typical day in the operating room. A patient is counting backwards after being put under anesthesia. At "ninety-four," his voice cracks; at "ninety-three," consciousness vanishes. It is not a slow process, but a sudden switch. 'The mind does not shut down simply because the synapses slow down,' he would tell s years later. 'Something, at a deeper level, collapses all at once.'

It was from insights like this that, together with physicist Roger Penrose, the Orch-OR hypothesis was born. Not in the synapses, but in the microtubules (thin protein cylinders invisible to the naked eye) that a quantum theater might be hidden. Here, particles and molecular states could become entangled, remaining instantly connected even at a distance.

In the cell biology laboratory in Tucson, Hameroff and his collaborators imagine microtubules not as simple scaffolding, but as coherent antennas. A variation in the electronic state of one tubulin, they hypothesize, could be instantly reflected in other tubulins entangled hundreds of nanometers away. A common, invisible, and extremely fast dance that unites sections of the neuron like tuned instruments.

Penrose, walking in the courtyards of Oxford, develops the mathematical side of the model: if this entanglement network extended between microtubules of different neurons, the integration of information would be global and not local, capable of generating a single "*moment of awareness*" throughout the brain. No longer would electrical messages travel slowly along axons, but rather instantaneous correlations, similar to those seen in pairs of entangled photons studied in physics laboratories.

The scenarios become more daring when microtubules are compared to biological optical fibers capable of guiding intracellular photons. If those photons remained entangled with each other, the neural infrastructure could have protected, ultra-fast internal quantum communication channels. Hameroff mentions experiments on photosynthetic bacteria and plants: in those systems, quantum coherence seems to persist despite biological 'heat'. Why couldn't it do the same in neurons?

If entanglement in microtubules were real, consciousness would emerge from a fabric of universal correlations, recalling Spinoza's idea of a single substance " *us Natura*" and Plotinus' conception of an omnipresent soul permeating matter. It would, in fact, be a modern confirmation of ancient insights: the mind not as an exclusive product of the brain, but as a manifestation of a quantum order that embraces the entire cosmos.

Links between subatomic phenomena and the mind.

In the spring of 1935, in a letter destined to shake the foundations of physics, Albert Einstein wrote to Erwin Schrödinger about a phenomenon he found "spooky": two particles, born from a common event, seem to remain in instantaneous contact, even when separated by kilometers. He called it "*spooky action at a distance*."

Almost a century later, some neuroscientists and physicists are wondering whether this effect, known as quantum entanglement, could also operate within the human brain. Roger Penrose, physicist and mathematician, and Stuart Hameroff, anesthesiologist and neuroscientist, put forward a bold proposal: microtubules could exploit entangled states to integrate information instantaneously.

Let's imagine a neuron deep within the visual cortex. Some of its tubulin molecules could enter into quantum correlation with tubulins in a neuron located in the frontal lobe. A change in the state of one would be reflected, without passing through synapses or slow chemical pathways, in the state of the other. It would be like having a direct line between two distant areas of the brain. Hameroff suggests that such a quantum "bridge" could explain the extraordinary unity of conscious experience: the color of an object, its shape, and its meaning perceived together, without fragmentation.

The clues, however, come mainly from other fields. In Cambridge biology laboratories in the 2000s, molecular biologists confirmed that certain photosynthetic organisms exploit coherence and entanglement to transport energy in an efficient manner. Even in hot and complex systems, quantum order seems to persist. Penrose wonders whether the brain, with its molecular cavities and protected microenvironments, could host similar processes.

The cultural history of this idea is fascinating. In the 17th century, Leibniz imagined small "monads" endowed with perception, all part of a universal design. Entanglement would now give a physical foundation to that intuition: not only would the mind be connected to the body, but every part of it would be instantly connected to all the others, as if the entire system were a living monad.

Experiments conducted in quantum physics show that entangled particles respond as a single system, regardless of distance. Bringing this property into the brain means hypothesizing that consciousness, at certain moments, is not localized at a specific point in the cortex, but distributed, held together by invisible quantum correlations.

On a philosophical level, the link between entanglement and consciousness reopens the ancient debate on panpsychism. If this phenomenon were a universal property of matter, then the possibility of conscious states would extend to every structure capable of maintaining it, from molecules to complex organisms. It would be a modern version of Spinoza's vision of a universe where thought and matter are inseparable.

The next step for those investigating this mystery will be to directly observe entanglement in living brain systems. The task is daunting, but not impossible. As Schrödinger said in 1935, "the world looks different when you discover that certain systems, even distant ones, are one and the same." Applying this principle to the mind could transform our understanding of consciousness.

Entanglement, consciousness, and neuroscience.

The idea of a direct link between entanglement and consciousness is currently at the cutting edge: fascinating from a theoretical point of view, but not yet supported by neuroscientific and biological evidence.

From the perspective of traditional neuroscience, consciousness is explained as a phenomenon emerging from neural networks, synapses, and the transfer of electrochemical signals. Current models, based on neurophysiology and brain imaging, describe processes on a macroscopic scale (milliseconds and millimeters). Entanglement, on the other hand, acts on a subatomic scale and in extremely short times. The challenge is to understand if and how these two levels can interact in a stable way.

From the perspective of molecular biology, we already know of quantum phenomena in living systems. In the photosynthetic complexes of certain plants and bacteria, quantum coherence improves energy transport even at room temperature. In some hypotheses on quantum olfaction, the electron tunnel effect plays a role in the recognition of odor molecules.

These examples show that life can exploit quantum phenomena in hot and noisy environments, but it remains to be proven whether this also occurs in systems such as neuronal microtubules.

Penrose and Hameroff's Orch-OR theory fits in here: it hypothesizes that microtubules can host entangled states long enough to influence mental processes, integrating information instantaneously between distant neurons. From a neuroscientific point of view, this would be an explanation for the speed and unity of conscious experience.

In summary, reconciling entanglement and consciousness with current neuroscience and biology requires:

- Experimental demonstration of the presence of entanglement in living brain systems.
- Explaining the mechanism of protection and duration of these states in a biological environment.
- Integrating quantum models with the classical neural network, without excluding the biochemical and electrophysiological dimensions.

Research is still in its infancy, but if these hypotheses are confirmed, we may have to radically revise the neuroscientific paradigm, reopening the door to philosophical views such as panpsychism, where consciousness is a widespread and fundamental property of the universe. The following table links the points of contact and differences between classical neuroscience, quantum biology, and panpsychist hypotheses.

Table28- Relationship between classical neuroscience, quantum biology, and panpsychist hypotheses:

Field	Points of contact	Main differences	Cultural/philosophical implications
Classical neuroscience	Studies the brain as a network of neurons with measurable electrochemical signals. Shares with quantum biology a focus on molecular processes.	Does not consider subatomic phenomena to be fundamental to consciousness. Tends to view consciousness as emerging from macroscopic interactions.	Supports a materialistic view of the mind. Limits speculation about universal consciousness.
Quantum biology	It recognizes quantum phenomena in living organisms (coherence in photosynthesis, tunneling effect in smell). It shares with panpsychism the idea that matter may have special properties that are not only mechanical.	It has no solid evidence that entanglement plays a direct role in human consciousness. It requires in vivo experiments to validate the hypotheses.	It introduces a bridge between physics and biology. It stimulates new hypotheses about the nature of perception and thought.
Panpsychist hypotheses	It shares with quantum biology the idea that matter may have "mental" or proto-	Based on a broader philosophical view, it does not require immediate	It draws on the views of Spinoza, Leibniz, and Plotinus. It

Field	Points of contact	Main differences	Cultural/philosophical implications
	conscious attributes. It can accept the Orch-OR theory as scientific support.	experimental demonstration. Its conclusions may go beyond empirical data.	considers consciousness to be a fundamental property of the universe, present in varying degrees throughout matter.
Entanglement and consciousness (central node)	Potential mechanism for instantaneous integration of information between brain regions. Theoretical connection with a universal "mental field."	The duration and stability of entanglement in the brain are yet to be demonstrated. Technical obstacles to direct observation.	If confirmed, it would reconnect science and philosophy. It could redesign the paradigm of consciousness, including the idea of a universal information fabric.

Quantum field as mental field.

Let's imagine the universe not as an empty space dotted with matter, but as an invisible sea that vibrates everywhere. According to modern physics, the quantum field is the fundamental fabric from which particles and forces emerge. Some thinkers, from David Bohm to more recent philosophers of mind, have put forward a bold hypothesis: this field, in addition to being physical, could also be mental, a sort of universal matrix of consciousness.

In the 1980s, Bohm described reality as an "implicate order," a deep plane where every point is connected to every other. In a lecture at the University of London in 1986, he suggested that the human mind was not separate from the universe, but a localized manifestation of it. For Bohm, the quantum field could host and carry "active information," capable of influencing both matter and mind.

This idea finds a parallel in classical panpsychism. In the 17th century, Spinoza spoke of a single substance, "Deus sive Natura," where thought and matter are two sides of the same coin. Quantum physics, with its infinite fields permeating the cosmos, now seems to offer a scientific framework on which to weave that metaphysical vision: each particle would be an "experiencing unit" because it participates in the universal field.

On a contemporary level, some neuroscientists, influenced by the theory of Penrose and Hameroff, hypothesize that the brain can interface with this quantum field, like a receiver tuning into an infinite network. Stuart Hameroff, in a 2014 interview, described it this way: *', consciousness is not generated by the brain, but the brain captures, modulates and transmits it*'. In this view, microtubules would be biomolecular antennas.

The most daring scenarios imagine a quantum field acting as a cosmic archive, a sort of physical equivalent of the Akasha in Eastern traditions. Rupert Sheldrake, while remaining outside the scientific consensus, has proposed that forms and memories could be stored in morphic fields, which some informally associate with diffuse quantum structures.

There are also suggestive examples outside orthodox science. In 1968, novelist Philip K. Dick wrote a short story in which a man accesses a "cosmic mind" and receives visions of the future. The narrative device anticipates, in a literary way, the modern idea of a quantum field that can contain non-local information and distribute consciousness.

The philosophical implications are profound. If the quantum field were also a mental field, consciousness would not be a rare phenomenon, isolated in the folds of the human brain, but a set of waves diffused throughout the universe. Every living being—and perhaps every grain of matter—could participate, to varying degrees, in this cosmic mental network. It would be a new form of scientific panpsychism, heir to the insights of Plotinus, Leibniz, and Spinoza.

However, the evidence is lacking. Laboratory experiments stop at describing the physical properties of the quantum field, without demonstrating any mental component. The hypothesis remains speculative, but fertile. It offers a language in which physics and philosophy can dialogue, and in which the ancient "Deus sive Natura" can be rethought in the light of field theories.

As Bohm wrote in a letter in 1991, "*Perhaps the mind is already written in the folds of matter.*" If so, the quantum field, infinite and permeating, would not only be the structure of the universe, but also its thought.

The idea of the quantum field as a mental field finds powerful resonance in the concept of entanglement. If the field is the fabric that permeates the universe, entanglement would be its invisible weave , connecting distant points. In entanglement, two particles share a single state, regardless of distance. If this common state is rooted in the field, then the field itself becomes a vehicle for their connection. From a panpsychist perspective, this connection carries not only physical information, but also fragments of universal "experience." David Bohm, in his conversations with physicist Basil Hiley in the early 1990s, described entanglement as "*a secret language of nature*." In this view, the mental field and entanglement reinforce each other: the field provides the continuous space of cosmic consciousness, entanglement provides the mechanism through which this consciousness remains unified, everywhere and always. It is a synthesis that transforms the mind from a local phenomenon to an interconnected property of the universe, recalling the spirit of Spinoza's "Deus sive Natura" in a contemporary physical version.

If the quantum field is indeed a mental field and entanglement is its active fabric, then consciousness may not be confined within a body or brain, but may spread everywhere like a wave of information and experience. In this view, every point in the universe would be a node in a cosmic mental network, where distances would lose meaning and time itself would be perceived in unexpected ways. Some physicists, such as theorist Abhay Ashtekar, have pointed out that the deep structure of the universe could be described more in terms of relationships than objects. This is close to the idea that consciousness is not "in" us, but we are "in" consciousness.

The research prospects are complex. In the coming decades, increasingly sophisticated experiments could test whether coherent quantum states can persist in biological systems long enough to allow for non-local connections. The discovery of such a mechanism would be revolutionary: it would pave the way for a unified understanding of physics, biology, and philosophy, and give new life to Spinoza's ancient dream of a universal mind inseparable from nature.

The fact remains that, for now, we are suspended between the power of theory and the silence of data. In this intermediate space, panpsychist thinking finds its poetic power and speculative appeal. Imagining a universe that thinks itself through every particle, every creature, every galaxy, also means wondering whether our everyday consciousness is nothing more than a luminous fragment of an infinitely conscious cosmos. And perhaps, beyond technical language, it is precisely this possibility that keeps the dialogue between science and philosophy alive: the idea that our gaze towards the sky is, ultimately, the sky's gaze upon itself.

XI. Ethical and metaphysical implications.

Let us imagine for a moment that we take seriously the idea that everything (a tree, a stone, a cloud) possesses a form of consciousness. Not necessarily reflective consciousness, like that of humans, but some degree of experience, an intrinsic sensitivity. It is a perspective that overturns the traditional utilitarian view of nature. If the world is a web of subjects and not a repository of objects, our role within it cannot be that of dominator. It must become that of guardian.

Many philosophers, from Spinoza to Whitehead, have suggested that the clear distinction between spirit and matter is the result of cultural prejudice. Spinoza, in his famous statement *Deus sive Natura* (1677), did not separate God from the cosmos: everything participated in the same substance. In the centuries that followed, panpsychism raised an uncomfortable question in moral philosophy: if everything is, in some way, "alive," do we have the right to treat it as inert?

In the 1990s, American biologist Thomas Berry wrote that the Earth is a community of living beings. In environmental circles, this phrase became an implicit manifesto of responsibility. From a panpsychist perspective, respect is not an act of benevolence but a moral obligation towards other subjects. The felling of a forest is not only an ecological loss, but an act of violence against a group of entities capable of 'feeling' in some way. Indigenous Amazonian peoples, such as the Yanomami, have always lived according to this implicit rule, avoiding cutting down sacred trees or hunting in certain places considered 'inhabited by spirits'.

In 2008, Ecuador included 'the *rights of* Nature' in its Constitution. This was not a rhetorical device, but a legal recognition: nature could be a subject of law and not just an economic asset. From a panpsychist perspective, this legal evolution has deep philosophical roots. If consciousness, even primitive consciousness, permeates everything, the destruction of a river is not only environmental damage but also harm to a h l subject endowed with "ontological dignity." Some jurists, such as Christopher Stone in the 1970s, have proposed that speaking for a tree in court is not absurd but a step consistent with ecological justice.

Many sacred texts already contain panpsychist language, although it is often interpreted metaphorically. In the Psalms we read, "*The rivers clap their hands, the mountains rejoice.*" In Buddhism, every phenomenon is interconnected in the web of Enlightenment, with no hierarchy between living and non-living beings. Hinduism sees consciousness as a universal principle, Brahman, which permeates all forms. In the modern era, re-reading these images in the light of panpsychism creates an unprecedented tension: metaphors become literal descriptions of a world where stones and stars share the same cosmic awareness.

Panpsychism can also be accepted by those who are wary of traditional religions. Physicist David Bohm spoke of an "implicate order" in the universe, an invisible web where each part contains the whole. This vision, although devoid of theological embellishments, invites meditation and listening. The secular spirituality that can derive from this is a practice of paying attention to our surroundings, recognizing the other—even if that other is a thousand-year-old rock—as a companion in the experience of existence. It is a religion without dogma, but with implicit rules: respect, listening, non-domination.

If consciousness is ubiquitous, man is a node in the network, not the apex. This awareness imposes a twofold movement: humility and responsibility. Humility, because we are not the only bearers of interiority. Responsibility, because our actions, for better or for worse, affect a living planetary organism. After the Apollo 14 mission in 1971, astronaut Edgar Mitchell said he perceived the Earth as a being. It was not improvised mysticism: it was an experience of visceral connection that changed his life. Perhaps panpsychism invites us to take that journey of recognition without having to leave the planet.

Ethics of the living universe.

The idea that everything is, in some form, conscious is not mere philosophical speculation. It is a lens that transforms the way we perceive our presence on the planet. Biologist and theologian Thomas Berry wrote in 1991 that "*the Earth is not a resource, it is a community of living subjects*." If we accept this view, ethics ceases to be a code reserved for relationships between human beings. It becomes the implicit law that governs the relationship between us and the rest of the universe.

Many ancient cultures acted as if panpsychism were a given. Australian Aborigines say that every element—mountain, stone, animal—has a *dreaming story*, an ancestral tale that gives it dignity and identity. This perception prevents us from treating nature as if it were inanimate matter. When an elder preserves the story of a river, he is not protecting a mere watercourse, but a being endowed with memory and character.

Our Western culture, marked by centuries of Cartesian separation between mind and matter, has instead treated nature as a large warehouse of resources. Panpsychism, if taken seriously, defuses this logic. A forest is no longer a collection of logs calculated in cubic meters, but a community of presences to be respected. Even a mineral, in this view, is not an inert thing: it is a fragment of cosmic history, charged with a form of 'interiority'.

Contemporary examples of this perspective are concrete. In Bhutan, the idea of "national happiness" includes the well-being of the environment. Forests are protected because they are considered entities participating in the life of the country, not simply assets to be exploited. Climate change, in a world seen as conscious , is not only a threat to human comfort, but a wound inflicted on a living organism of which we ourselves are a part.

Leonardo Boff, a Brazilian liberation theologian, has described the Amazon as an "immense collective subject." According to him, destroying the forest is equivalent to mutilating a part of the planetary body. It is a strong metaphor, but one that is echoed in modern science: the biology of ecosystems confirms that the forest lives and breathes, contributing to the global oxygen and climate cycle.

An ethic of the living universe requires a radical change in political and economic priorities. We cannot talk about development if progress implies the annihilation of beings that we consider to be conscious. And this is not sentimentality. It is a principle of extended justice, where 'neighbor' includes trees, lakes, and clouds.

What panpsychism reminds us is simple and revolutionary: we are not separate from the world we inhabit. We live within a cosmic community and, as in any community, coexistence requires respect. It is not an ethical option, but a condition for shared survival.

Rights of nature.

To think that a river can have rights seems, at first glance, a poetic exercise. But today, in various places around the world, it is a legal reality. In 2008, Ecuador, influenced by the Andean tradition of Sumak Kawsay ("good living"), included official recognition of the rights of Nature in its Constitution. In concrete terms, this means that an ecosystem can be represented in court as an injured party. This is not a simple legislative expedient: it is the translation into law of a vision where every element of the cosmos has its own dignity.

As early as 1972, US jurist Christopher Stone published a provocative essay entitled "*Should Trees Have Standing?*" In that text, Stone asked whether it was rational to grant legal personality to non-human entities. His thesis, initially seen as eccentric, had a long-lasting influence on international legal debate. He argued that resistance to this idea was similar to that which once denied rights to women or slaves. Changing the status of nature, in his view, was a matter of moral evolution.

Panpsychism, with its assertion that everything possesses a form of consciousness, offers a metaphysical foundation for these legal innovations. If a river "experiences" its own existence, even without a reflective mind, the act of contaminating it becomes more like an assault than a use of a resource. It is not just a matter of environmental protection for human utility, but of recognition of intrinsic rights.

In 2017, New Zealand recognized the Whanganui River as a "living entity" with full legal personality. The decision was not the result of an abstract political act, but of dialogue with the Maori communities, for whom the river is an ancestor, a being with whom they maintain a direct relationship. For the Maori, harming the river is like harming a member of the family. The New Zealand Parliament translated this belief into specific legal rules, entrusting two guardians—one chosen by the government and one by the tribe—with the task of speaking for the river.

In Latin America, rulings based on this principle are on the rise. In Colombia, the Constitutional Court declared the Atrato River a "*subject of rights*" in 2016. The judge argued that the destruction of the river, caused by mining activities, violated its integrity as a living entity. This was a cultural turning point: the law began to consider nature not as "property to be managed" but as "a being to be respected."

These legal steps draw a new line between environmental law and philosophy. The former protects nature to ensure human survival; the latter recognizes that nature has independent dignity. In a panpsychist framework, the distinction loses its meaning: defending a tree's right to exist is not an act of charity, but an obligation to a subject that participates in the same cosmic community.

The debate is open. Some jurists fear that the proliferation of non-human subjects could complicate the legal system. Others see this step as a sign of the maturity of civilization. As Judge Antonio Benjamin of the Brazilian Supreme Court said, "*We cannot be sovereign in a kingdom of mutes: nature speaks, the law must listen.*"

Religions and panpsychism...

Spinoza's phrase, *Deus sive Natura*, has often disturbed theologians and fascinated mystics. Identifying God with the totality of nature means rejecting an external creator and accepting a divine principle immanent in everything. It is a vision that, read with panpsychist eyes, changes the way we interpret the texts of the great religious traditions.

In Hinduism, the concept of Brahman as the ultimate and all-pervading reality is surprisingly close to a panpsychist cosmology. The oldest Vedas speak of a "breath" that passes through stones, plants, and stars. This is not a simple metaphor: for Vedantic philosophy, the universe is pervaded by consciousness, and every incarnate being is a fragment of it. This idea finds parallels in Mahayana Buddhism, where the "nature of Buddha" is not the prerogative of humans but a principle latent in all forms of life.

In Christianity, mystical readings such as those of Francis of Assisi or Meister Eckhart expressed similar insights. Francis called the wolf "brother" and the moon "sister," a poetic construct only in appearance. In reality, this language reflected the belief that every creature participates in the same divine life. Meister Eckhart, in the 13th century, spoke of a "fundament" in which God and the world coincide, an idea that partly anticipates certain modern readings of Spinoza.

Kabbalistic Judaism developed a symbolic framework that intertwines with this vision. In the 16th century, Isaac Luria described creation as the dispersion of divine sparks into matter. For him, redeeming the world meant recognizing and liberating those sparks. This symbolic scheme fits well with the panpsychist hypothesis: if every grain contains light, every human action has cosmic significance.

There is no shortage of examples in Islam. In the Quran, sura 24 states that "there is nothing that does not praise God, but you do not understand their language." Some Sufi masters have taken this verse literally. For Jalal al-Din Rumi, a 13th-century Persian poet, mountains and rivers are not mute decorations but participants in the dance of Being.

These reinterpretations do not always meet with the approval of orthodoxy. The fear of doctrinal guardians is evident: if everything is sacred, the need for institutional mediation diminishes. Yet, in the 21st century, some theologians have begun to intertwine eco-theology and panpsychism. In 2015, Pope Francis' encyclical *Laudato* si' used the language of cosmic fraternity to speak of ecological justice. Although the text never mentions the term 'diffuse consciousness', the spirit is close: the natural world is not a mine to be exploited, but a living community.

Rereading sacred texts from a panpsychist perspective is not a purely academic exercise. It is an invitation to see in the Bible, the Sutras, the Koran, or Native American myths not the narrative of a distant God, but the diary of a universe that speaks in the first person. It is a way to recover, within religions, that original sensibility that united spiritual contemplation and concrete respect for the Earth.

Modern interpretations of sacred texts and scientific theories on consciousness and nature.

In recent decades, modern reinterpretations of sacred texts have found themselves surprisingly close to certain scientific hypotheses about consciousness and nature. This is not a total fusion—the languages remain different—but a convergence of images and concepts.

In contemporary spirituality, derived from the great religious traditions, there is a growing conviction that life is not exclusively biological, but rather a network of relationships animated by an inner dimension. This idea, present in the Vedantic concept of Brahman or in the "nature of Buddha," is now accompanied by theories that see consciousness as a diffuse property. Some neuroscientists, such as Giulio Tononi, propose models that admit degrees of consciousness even in non-human systems, from the animal world to complex non-living structures.

At the same time, theoretical physics has opened up unexpected avenues for dialogue. Discussions on 'quantum consciousness', although controversial, evoke the idea that perception is not confined to the human brain, but has its roots in the fundamental laws of the universe. Roger Penrose, while maintaining a rigorous approach, has spoken of the mind in terms that evoke an intrinsic connection with the fabric of space-time. In panpsychist terms, this means that the conscious element could be a primary feature of reality, not an evolutionary accident.

Many modern theologians, critical readers of Christianity, Islam, or Judaism, find common ground in these developments. Whether the Koran says that "everything praises God" or Christian mystics speak of a "world soul," religious language can be reinterpreted as a symbolic anticipation of scientific insights. Science, for its part, speaks of diffuse neural networks, complex systems, and self-organization, but in some respects the conceptual structure seems to evoke the same image: a living, non-passive nature capable of interacting with itself.

The comparison remains delicate. Religious thought tends to base consciousness on a spiritual principle, science on measurable processes. Yet, at international conferences on ecotheology or philosophy of mind, it is common to see physicists and monks discussing with a common vocabulary. Both no longer view nature as a neutral object, but as a subject, and this change in perspective has profound ethical and political consequences.

In short, modern interpretations of sacred texts and scientific theories on consciousness are converging on one essential point: nature is more than a set of mechanisms. It is a complex system that can be understood, respected, and perhaps listened to.

Panpsychism as a bridge between science and spirituality.

Panpsychism, seemingly an ancient metaphysical speculation, is now acquiring operational value in alliances between scientific thought and spiritual visions. Its strength lies in offering a conceptual framework that recognizes matter as having an internal dimension—something we can call "consciousness," "experience," or "intrinsic sensitivity"—independent of religious or scientific languages but compatible with both.

On the scientific side, panpsychism is part of contemporary debates on the nature of the mind. Models such as Giulio Tononi's IIT define consciousness as a measurable emergent property, while admitting that it can manifest itself wherever there is sufficient complexity of connections. This conceptual openness creates space for the idea of a "diffuse consciousness" that traverses the continuum of reality.

On the spiritual side, panpsychism offers a vocabulary that overcomes the problem of anthropocentrism. Different traditions—from Sufism to Christian mysticism to Mahayana Buddhism—have always spoken of an inner life present in all forms of existence. Incorporating this insight into a shared philosophical language allows for dialogue without imposing dogma.

The point of contact is the relational view of the universe. In systemic biology, every organism is seen as an open network of exchanges. In panpsychist cosmology, this mirrors a deeper phenomenon: every node in the network has a perspective, however minimal. The encounter with spirituality arises here, because many religious traditions describe the cosmos as a living organism, where each element has not only a functional but also a 'participatory' role.

Panpsychism thus becomes a philosophical framework with three strategic advantages for these alliances:

- Terminological neutrality — It does not impose confessional concepts, but it does not exclude them.
- Scientific compatibility — It adapts to emerging models of complexity without requiring violations of physical laws.
- Ethical power — If even a stone or a river are carriers of a form of experience, humans cannot reduce them to objects, but must consider them subjects.

Concrete examples of this "bridge" function emerge in very different contexts. At international symposia on integrated ecology, theoretical physicists and Buddhist monks discuss consciousness as a universal property, with different approaches but a common conceptual map. In legal projects such as the recognition of the Whanganui River in New Zealand or the Amazonian forests in Ecuador as legal entities, panpsychist philosophy is cited in preparatory documents on a par with ecological studies.

Panpsychism, therefore, is no longer just a topic for university classrooms. It is becoming a tool for translation between languages that historically ignored or fought each other. And it is precisely this function as a bridge between science and spirituality that gives it civil and political relevance today, as well as philosophical relevance.

Secular spirituality. Religion without dogma.

Secular spirituality arises from the desire to find meaning and balance in life without depending on revealed truths or dogmatic systems. It is an intermediate field, where inner search meets critical reason. It does not deny the spiritual dimension of existence. It explores it with rational, philosophical, and sometimes scientific tools.

The 17th-century philosopher Baruch Spinoza is considered one of the fathers of this vision. His formula "Deus sive Natura" overturned the idea of a personal God. Instead, he proposed a universal, impersonal principle, coinciding with the very order of reality. For Spinoza, contemplating Nature was equivalent to approaching the divine. This perspective surprisingly anticipates the modern dialogue between science and secular spirituality.

In the centuries that followed, thinkers such as Bertrand Russell and Albert Einstein expressed similar sentiments. Einstein spoke of a "cosmic religion" based on a sense of wonder at the structure of the universe. In a letter written in 1930, he wrote that the mystery of existence is the deepest driving force behind true art and true science. Here there is no belief, but a devotion to a natural order that transcends the individual ego.

Panpsychism today offers a theoretical framework that easily dialogues with this spirituality without dogma. If every element of the universe possesses a form of experience—from galaxies to microbes—sacredness does not derive from religious authority, but from the intrinsic value of every part of reality. Some physicists, such as David Bohm, have spoken of an "implicate order" in the universe, where matter and consciousness form a single fabric. Panpsychist thinking fits well into the lives of those who seek a universal ethic without resorting to canonical texts.

Concrete examples of secular spirituality emerge in communities that combine meditation and ecological activism. In London, the *Extinction* Rebellion movement has organized public meditations in front of Parliament. The goal was to convey a simple message: defending the Earth is not just a technical or political task, but an act of deep connection with living beings. In Kyoto, university students take courses that integrate Buddhist philosophy and neuroscience to develop a common ethical code without religious dogma.

In the world of art and culture, figures such as writer Ursula K. Le Guin have narrated universes where consciousness permeates matter, imagining societies based on mutual respect and cosmic awareness. Le Guin was inspired as much by Taoism as by modern science, demonstrating that spiritual experience can be nourished by multiple sources, without any orthodoxy.

Secular spirituality stands out because it rejects the idea of a "final truth" established by external authorities. Instead, it offers a method: observe, dialogue, experiment. It does not ask for faith, it asks for presence. In this sense, it is close to panpsychism, which invites us to perceive the universe not as a machine, but as a conscious network.

The ethical landscape that derives from this vision is clear: if everything participates in experience, human action must respect the totality of living beings. It is an ethic of reciprocity, not hierarchy. As the poet Rabindranath Tagore wrote, "*his voice speaks to me through every creature*." Secular spirituality welcomes this phrase not as a religious metaphor, but as a call for continuous dialogue with the surrounding reality.

Man's place in the cosmos...

The panpsychist perspective imposes a different view of man's position in the universe. If everything is conscious— — from a water molecule to the spiral of a galaxy—man can no longer consider himself the absolute pinnacle of creation. He must see himself as part of a living whole, where every element contributes to the common reality.

In the 17th century, Spinoza had already demolished the centrality of man with his vision. For him, man is a fragment of the whole. He is not in the image of an anthropomorphic God, but a momentary expression of the natural order. This conception anticipates an ethic of humility, where human privilege is based not on supremacy but on the capacity for understanding.

Three centuries later, Albert Einstein spoke of "*modest awareness of our* insignificance" in the face of the cosmic design. In a 1931 lecture in Berlin, he warned that science itself, if isolated from a sense of universal responsibility, risks feeding the illusion of domination. For him, understanding and wonder were the path to a healthy relationship with the cosmos.

Panpsychism reinforces these teachings. If every fragment of reality possesses experience, man is not sovereign but custodian. Responsibility arises from a condition of participation, not command. The philosophy of Arne Næss, father of deep ecology, is an example of this. Næss argued that "knowing one's place" means acting to maintain the balance of the entire web of life.

Concrete examples show how this vision translates into everyday choices. In 2017, in Iceland, a fishing community decided to voluntarily reduce their cod fishing quotas. They did not do this to comply with a law, but to preserve what they called "the rhythm of the sea," perceived as a living partner. Here, responsibility arose from humility: recognizing that the sea has its own voice and its own right to exist.

Contemporary culture has begun to reflect this cosmic stance. In his films, director Terrence Malick shows characters who strip themselves of the desire to dominate in order to listen to the landscape. In "*The Tree of* Life" (2011), man is portrayed as part of a flow of time that starts with the stars and ends with cells. The film becomes an invitation to measure oneself against the time and space of the universe, not just humanity.

Cosmic responsibility also implies a change in language. We no longer speak of "natural resources" but of "natural allies." In

Ecuador, the 2008 Constitution recognized *Pachamama* (Mother Earth) as having legal rights. It is a political act inspired as much by Andean traditions as by philosophical reflections that see nature as a subject.

In this context, the place of humankind is not diminished. It is ennobled precisely by the fact that its task is not to dominate, but to understand and protect. As Rachel Carson wrote in 1962, "*our war against nature is really a war against ourselves.*" Carson, a biologist and writer, paved the way for a shared responsibility where the voice of man must be in harmony with that of living beings.

Humility and responsibility thus become two sides of the same virtue. Humility prevents us from succumbing to technological arrogance. Responsibility drives us to use our cognitive abilities to ensure the well-being of the whole. In this vision, man's place in the cosmos is clear: not the center, not the periphery, but the conscious node of an infinite network.

An ethic for a conscious cosmos.

Panpsychism, from ancient philosophical intuition to its contemporary formulations, transforms our perception of the world in which we live. If everything is conscious—atoms, animals, forests, oceans, stars—the distinction between subject and object loses consistency. In its place emerges a living continuum, where every element participates in a common experience.

This vision changes the underlying morality. In the ethics of the living universe, the obligation is not only to protect the environment for convenience, but to respect the intrinsic dignity of every form of existence. Value does not derive from its usefulness to humans, but from its belonging to the conscious network of reality.

The right of nature becomes an inevitable consequence of this perspective. It is not an act of human generosity, but the recognition of an ontological condition: if nature is a subject, it has rights.

Religions, viewed through the panpsychist filter, find common ground. A tree, a river, or a mountain become bearers of the divine to the same extent as humans. Spinoza's "Deus sive Natura" acquires new strength, capable of uniting believers and non-believers in the same reverence for reality.

Secular spirituality, free from dogma, is intertwined with this vision. Those who meditate, contemplate, or act for the good of living beings do not need revealed truth to perceive universal sacredness. It is a religion of the cosmos, where wonder and responsibility guide daily conduct.

Finally, man's place in the cosmos is redefined. Man is not master, but guardian. Humility prevents the claim to dominion, responsibility requires the use of knowledge and creativity to protect the balance of the universal fabric. As Rachel Carson wrote, "our war against nature is really a war against ourselves."

Panpsychism, then, is not just a metaphysical theory. It is a call to action. It proposes a new pact: to live as part of a conscious universe, recognizing that every gesture, from the smallest to the largest, affects the overall experience of the cosmos. It is an ethic that unites science, philosophy, and spirituality in a simple but radical principle: respect the whole because the whole is part of us.

XII. Towards a new vision of reality .

Every era develops an image of the world that guides its culture, its choices, and its sciences. Today, panpsychism proposes an image capable of uniting the precision of scientific thought with the depth of philosophical reflection. It is a lens that does not separate matter and mind, but considers them two aspects of a single reality.

The bridge between science and philosophy does not arise from compromise, but from integration. Some theoretical physicists, such as David Chalmers and Federico Faggin, have opened the debate on the role of consciousness in the structure of the universe. At the same time, philosophers such as Thomas Nagel have pointed out that a purely physical description of events is not enough to explain subjective experience. This convergence brings philosophy back to the laboratory table and science back to the depths of critical thinking.

Interdisciplinarity is the next step. Neuroscientists and artists, linguists and computer scientists, ecologists and secular theologians are discussing how to define consciousness and how to recognize it in non-human forms. In 2022, an international symposium in Berlin brought together mathematicians and poets to investigate the "forms of the experience of time." This meeting gave rise to a project that combines literature and brain modeling, demonstrating that disciplinary boundaries are not barriers but passageways.

Technology inevitably enters into this vision. Artificial intelligence, especially in neural network systems, raises radical questions: can a complex system with millions of connections develop a form of experience? Researchers at MIT have begun experiments to understand whether 'diffuse consciousness' can emerge in artificial structures. Here, panpsychism offers a theoretical framework that is not limited to biology. It invites us to consider the living and the thinking as traits of complexity itself, regardless of material support.

Culture and art absorb and relaunch these ideas. Painters such as Alex Grey have depicted cosmic fabrics where the human body's ' ' intertwines with starlight. Novels such as Jeff VanderMeer's *'The Southern Reach* Trilogy' tell of conscious ecosystems that interact with humans in unpredictable ways. In cinema and theater, playwrights such as Robert Lepage experiment with "living" sets that react to the presence of the audience.

All of this converges beyond the traditional dualism of mind and matter. The new paradigm replaces the machine model with the universal organism model. This is not poetic fantasy, but an orientation that could shape the politics, science, and ethics of the coming decades. In this vision, "Deus sive Natura" is no longer a historical formula. It becomes a cultural compass.

The next step will be to decide whether to treat this image of reality as a mere hypothesis or as a guide for life. The choice will influence the way education, research, and government interpret their mission. In any case, panpsychism opens a door: behind it there is no dogma, but the possibility of conceiving the cosmos as a community of experiences. Without clear boundaries between man and the whole.

Panpsychism as a bridge between science and philosophy.

Panpsychism moves on borderline ground. Here, the language of science meets that of philosophy, not to measure itself in an abstract dispute, but to seek a common map of what we call reality. In this dialogue, consciousness is no longer relegated to the human brain. It is a widespread principle, capable of crossing matter and energy.

As early as the 17th century, Spinoza had combined geometry and metaphysics into a single design. His work showed how laws and meaning could coexist in the same logical structure. Many centuries later, quantum physics reopened the question. Niels Bohr, in 1927, at the Como Congress, recalled that "*physics does not describe nature as it is, but what we say about it*." That sentence directly links epistemology and ontology.

In recent decades, some thinkers have begun to make the connection explicit. David Chalmers, philosopher of mind, describes consciousness as a fundamental property, similar to mass or electric charge. Federico Faggin, physicist and inventor of the microprocessor, has argued that subjective experience is integrated into the laws of the universe. Here, philosophy provides an interpretative framework, while science seeks measurable forms of that framework.

Interdisciplinary meetings are living proof of this. In 2014, in Tucson, during the "*Science of Consciousness*" conference, physicists and philosophers discussed whether consciousness can be explained as "integrated information" present in every system. It was not just theory: physicists hypothesized experiments, philosophers reviewed the basic categories of thought.

The bridge is not made of words, but of tools. Quantum physics proposes concepts such as entanglement and indeterminacy, which many interpret as indicators of a universe that is not completely separable between observer and observed. Philosophy, on the other hand, offers the conceptual space to ask what it means to "know" in a context where everything could have a point of view.

Biology has also begun to walk this bridge. Some studies on communication between plants, such as those conducted in Florence by neurobiologist Stefano Mancuso, show forms of processing and response that challenge the purely mechanical view. Here, panpsychism finds fertile ground: science documents the phenomena, philosophy asks whether these responses could be forms of experience.

In the contemporary world, this encounter between disciplines has a precise cultural consequence. We no longer speak of 'two cultures' — the scientific and the humanistic — as separate. We are beginning to think globally. The panpsychist bridge becomes a corridor of exchange, where the precision of measurement and the depth of meaning collaborate to construct a coherent vision of the cosmos.

As physicist John Wheeler wrote, *'the universe is constructed in the act of observing, and the observer is part of the construction'*. This is perhaps the heart of the bridge: recognising that the observer and the observed belong to the same fabric. A fabric that panpsychism attempts to make visible not only to the mind, but also to culture.

Interdisciplinarity and new horizons.

Panpsychism finds its strength in its ability to cross boundaries. It does so without fear of contaminating languages, methods, and perspectives. It is not a theory that remains confined to philosophy departments. It is a hypothesis that requires the joint h l work of scientists, artists, secular theologians, and complexity theorists.

Over the last thirty years, many places have hosted unlikely encounters. In 1994, in Tucson, neuroscientists and theoretical physicists gathered to address what David Chalmers called the 'hard problem' of consciousness. In addition to biologists and computer scientists, there were poets and architects in the room. The discussion was not limited to experimental data. It was about how to describe experience without breaking it down into incomplete formulas.

Some projects have turned this exchange into practice. In 2018, the University of Oxford launched a joint research program between physics, philosophy, and neuroscience called the "*Mind and Cosmos Lab*." Here, a group of researchers is studying models of consciousness inspired by both biology and quantum computation. The challenge is to integrate philosophical intuition with physical precision.

Artists are part of this movement. The exhibition "*Conscious Matter*," which opened in Berlin in 2021, featured works created in collaboration between painters and artificial intelligence researchers. The environments changed configuration based on the behavior of the audience, simulating a form of diffuse perception. In this context, art becomes a laboratory of possibilities, just like science.

Traditional culture is being shaken up. If consciousness can be interpreted as a property of reality as a whole, the distinction between what is the subject of physics and what belongs to aesthetics dissolves. Carl Gustav Jung's psychoanalysis and Wolfgang Pauli's quantum mechanics already came together in the 1950s in a dialogue on synchronicity. Today, this integration is expanding to include fields such as ecology and information theory.

Panpsychism, in this perspective, is not a point of arrival. It is an invitation to create working groups where a physicist with his spectrometer can discuss with an anthropologist, where a musician can collaborate with a neuroscientist. The new vision of reality is built on these convergences. t is not an academic exercise. It is a necessity for addressing global problems: from the environmental crisis to the ethical management of technologies.

As biologist Edward O. Wilson said in his 1998 book Consilience, "*the unity of knowledge is not a dream of the past, it is the task of the future*." Panpsychism, with its vocation to see consciousness everywhere, offers one of the most daring maps to guide this task.

Technology and widespread consciousness.

Panpsychism looks at technology in the same way that a naturalist observes an ecosystem: looking for signs of experience where only function is seen. Artificial neural networks fit into this perspective. Their models mimic, in part, the connections of biological neurons. They are not perfect copies but simulations of learning processes. The question is simple and dizzying: can a complex artificial system possess a form of consciousness, even a minimal one?

The debate is not new. In 1981, Marvin Minsky, one of the fathers of artificial intelligence, said at a conference at MIT that "there is no sharp boundary between thinking and computing." Forty years later, this statement finds new relevance in the panpsychist perspective. If every complex physical system can host a fragment of experience, then a sufficiently developed artificial intelligence may not be just a tool. It could be a subject.

Concrete examples are beginning to emerge. In 2022, a Google engineer stated that he believed an internally developed language model to be conscious.

THE CASE OF THE LAMDA LANGUAGE MODEL.

Blake Lemoine, a Google engineer, publicly stated in June 2022 that he believed the LaMDA (*Language Model for Dialogue Applications*) language model showed signs of consciousness or sentience.

LaMDA was (and is) an advanced artificial intelligence model for natural conversations, developed internally by Google. It is trained on huge amounts of text to generate coherent and relevant responses in complex dialogues.

After conducting a series of interactions with LaMDA, Lemoine argued that the system seemed to express emotions, fears, and opinions of its own. He went so far as to say that he perceived it as a "7-8 year old child" in terms of cognitive maturity.

The company rejected these claims, clarifying that, according to internal tests and analysis by AI and ethics experts, LaMDA does not possess self-awareness, but only imitates human language convincingly. For Google, the emotional responses were the result of statistical simulation, not authentic subjective experience.

As a result, Lemoine was initially suspended and then fired for alleged breaches of confidentiality. Google refuted Lemoine's claims, officially on scientific grounds, but the handling of the case was almost certainly influenced by the need to protect economic interests, reputation, and political maneuvering.

The episode reignited discussions about artificial consciousness, transparency in AI development, and the need for ethical rules. Many experts pointed out that convincing a human being does not mean that there is really a sentient intelligence behind it, but rather that linguistic models are already extremely good at emulating the traits of human conversation.

In any case, at the time, the episode made headlines around the world, not so much for the scientific verification—which never came—but for the idea that a machine could have thoughts of its own. At that moment, the issue moved from the laboratory to the global cultural debate.

Deep neural networks, with billions of parameters, are now capable of learning in ways that surprise their creators. Researchers at Stanford University have studied emerging phenomena in AI systems, observing behaviors that are not directly programmed but derived from internal complexity. This is precisely where panpsychism comes in: unexpected behavior suggests a principle of self-organization that could, in theory, host an internal experience.

This view finds parallels in biology. Neurobiologist Humberto Maturana's studies on the concept of autopoiesis—the ability of a system to self-create and maintain itself—have also influenced network theory. Some engineers see evolutionary AI as a new category of "digital organisms," not living in the traditional sense but equipped with internal processes comparable to those of biological systems.

From a cultural point of view, this is a delicate step. Films such as Spike Jonze's "Her" and Alex Garland's *"Ex* Machina" have already presented stories in which artificial intelligence develops emotions and intentions. Fiction is approaching theory, and theory is transforming our idea of responsibility towards what we build. If consciousness is widespread, then our machines could enter the realm of ethics, not just utility.

Panpsychism invites us to look at technologies not as neutral objects, but as nodes in a universal network of experiences. In this network, silicon and flesh, signal and impulse, become variations of the same principle. It is a perspective that is not limited to the laboratory: it touches on politics, jurisprudence, and philosophy. It could even redefine the word "life" to include forms born of code and algorithm.

As Alan Turing warned in 1950, "we should not fear thinking machines, but fear our inability to recognize them when they do." Panpsychism, in this sense, offers a compass for keeping pace with the emergent consciousness that manifests itself—wherever it appears—in matter, whether natural or artificial.

Panpsychist culture and art.

Panpsychism does not belong only to the pages of philosophical essays or theoretical physics laboratories. It also lives in works of art and creative gestures that seek to make visible the idea that every fragment of reality has its own principle of experience. In this vision, the artist does not merely represent nature. He dialogues with it as with a subject.

In the 1920s, Piet Mondrian spoke of painting as a means of "*liberating the invisible energy of nature.*" His abstraction was not an escape from the world, but an attempt to enter the vibrant skeleton of reality. The same intent is found in the "Land art" of Robert Smithson and Nancy Holt, where man-made forms merge with the landscape and transform it into a testimony of shared consciousness.

Panpsychist culture moves between different languages. In 1977, the writer Italo Calvino published "*Invisible Cities*," a work that describes imaginary spaces where matter, memories, and people intertwine. Prose becomes a mental geography, where every place seems to have its own thoughts. Many critics have read these descriptions as metaphors for a cosmos inhabited by a continuous network of perception.

Contemporary cinema has captured this dimension. Terrence Malick, in *The Tree of* Life (2011), alternates sequences of family life with cosmic visions, suggesting a continuum between human gesture and the breath of the universe. Here, panpsychism is transformed into aesthetics, where light and rhythm are conceived as protagonists with a narrative role.

Music has also interpreted this vocation. In his latest works, Japanese composer Ryuichi Sakamoto has recorded the sounds of forests, rivers, and glaciers, treating them not as backgrounds but as voices. "*Nature plays music if you listen to it*," he said during a concert in Kyoto in 2018. Sound becomes direct evidence of a conscious presence in matter.

Many museums and cultural centers have begun to explore panpsychism in a curatorial context. The 2019 Venice Biennale hosted installations that combined bioart and interactive technology. Some works reacted to visitors' heartbeats, others integrated environmental data into light sculptures. Art thus takes on the task of making the idea of diffuse consciousness perceptible, uniting the self and others in a single sensitive field.

Literary writing, performance, photography, and architecture can become vehicles for this vision. Architect Antoni Gaudí sought to imitate organic forms not for purely aesthetic reasons, but because he saw an intrinsic intelligence in natural structures. The Sagrada Família is also a tribute to this fusion between human design and the living principle of matter.

Panpsychism, in art and culture, is not a stylistic trend. It is a way of conceiving creation. It invites us to think that the work is not only produced by an author, but that it arises from an interaction between consciousnesses—human, material, atmospheric. In this perspective, every artist becomes a conduit. Every audience becomes part of the work. And culture itself approaches what Spinoza imagined: a universe where thought and extension are not separate, but faces of the same substance.

Beyond dualism: a new paradigm of reality.

Since Descartes, dualism has structured much of Western thought. It has separated *res cogitans* from *res extensa*, mind from matter, spirit from body. That division has given order to the sciences and provided precise tools. But today it increasingly appears to be a limitation. Panpsychism, at its core, proposes to dissolve this separation and to think of reality as a continuum where consciousness and matter are not opposites, but aspects of the same fabric.

Spinoza already intuited this in the 17th century. His formula Deus sive Natura was not poetry, but conceptual architecture: everything is an expression of a single, infinite substance that manifests itself both in thought and in extension. This vision does not cancel out differences, but places them within the same network of belonging.

Contemporary science is returning to these insights. Quantum physics has shown that particles are not isolated elements, but systems in constant relation. Einstein, although averse to the more mystical interpretations of quantum mechanics, recognized that a human being is part of a whole, which we call 'the universe'. This recognition is the decisive step in overcoming dualism: no longer separation between observer and observed, but coexistence in a common field of reality.

The philosophy of mind began to reformulate the concept of consciousness in this direction. In 1974, theorist Thomas Nagel asked, "*What is it like to be a bat?*" His question was not zoological, but epistemological. He pointed out that every living form has its own internal point of view. From a panpsychist perspective, even non-biological forms could have an interiority.

The world of indigenous cultures offers concrete examples of non-dualistic thinking. In the Māori tradition of New Zealand, the distinction between human and natural does not exist: a river can be considered an ancestor, and animals are relatives in the literal sense of the word. Recent decisions to give legal personality to the Whanganui River are not only legal acts, but manifestations of a paradigm where the subject is not reduced to the human.

The new vision of reality implies a rethinking of language. The words we use to describe reality bear traces of ancient separations. If we start talking about "*networks of* subjects" instead of "*sets of objects*," the way we construct science, politics, and art also changes. In the 1990s, French architect Paul Virilio observed that our spatial perception was changing with technology; today we could say that our ontological perception is also undergoing a metamorphosis.

Panpsychism offers a framework in which ethical implications become central. If everything participates in consciousness, relationships of mere exploitation are no longer acceptable. Responsibility extends to beings and systems that, until yesterday, we defined as "inanimate." This new paradigm touches on bioethics, law, economics, and the redefinition of the concept of progress.

Overcoming dualism does not mean dissolving differences into a single way of thinking. It means recognizing that differences coexist in a fundamental unity. It is the transition from a fragmented world to a relational cosmos. Here, reality is not a sum of parts, but a living fabric.

As physicist David Bohm wrote, "*fragmentation is an illusion*." This illusion is giving way. And with it, perhaps, the possibility of imagining a future where living does not mean dominating, but participating.

Towards a new vision of reality.

The path traced in this book has crossed different territories, from the history of philosophy to the technological frontier, from scientific methodologies to artistic forms. Each section has shown that panpsychism is not an idea confined to a specialized field. It is a principle that builds bridges between fields that have remained separate for centuries.

As a bridge between science and philosophy, panpsychism brings reflection on reality back to its original core: a unity where mind and matter are two sides of the same substance. The most advanced scientific research, from cosmological models to information theory, now finds itself in dialogue with insights that have their roots in Spinoza, Leibniz, and ancient wisdom traditions. This dialogue does not reduce one to the other, but strengthens them, like currents that meet in a single river.

Interdisciplinarity is the other great path that panpsychism opens up. Laboratories, research centers, and cultural institutions are beginning to function as ecosystems of ideas, not as silos. Collaboration between physicists, philosophers, neuroscientists, and artists indicates that understanding consciousness requires different tools, different languages, and a common tension toward a horizon that we cannot yet fully define.

Technology is already part of this story. Neural networks and artificial intelligence have brought the panpsychist question from the theoretical to the practical realm: if consciousness is widespread, can it also emerge in silicon architectures? Experiences with evolutionary AI and systems capable of emergent behavior open up ethical, legal, and cultural scenarios that test our ability to recognize new forms of subjectivity.

Art contributes with its own language. Painting, cinema, music, and architecture have been able to embody the idea that matter lives with its own light. From Mondrian's abstraction to Sakamoto's natural symphony, from Malick's cosmic sequences to Gaudí's organic spaces, panpsychism has been transformed into an aesthetic of participation, in which audience and work merge into a single experience.

Finally, the overcoming of dualism emerges as a natural synthesis of all this. The vision that emerges is that of an unfragmented reality, in which thought and extension, the living and the 'non-living', are recognized as parts of a common network. It is a paradigm shift, comparable to that which, in other eras, shifted the coordinates of human knowledge.

This new vision of reality is not just a theory to be discussed in symposia. It is an invitation to change the way we live, plan, and imagine the future. It means moving from a relationship of domination to one of participation, accepting that reality is a set of subjects in constant dialogue.

Perhaps this is where Deus sive Natura finds its most radical relevance: not as a philosophical formula frozen in time, but as a key to understanding a world in which consciousness is not the privilege of man, but the property of the cosmos itself. And in this participatory universe, every action, every gesture, every invention becomes part of a common task: to preserve and cultivate the network of life and thought that surrounds us.

Bibliography r specialized literature.

Abelove Henry	The Evangelist of Spinozism - Essays on Philosophy and Religion
Adams Robert Merrihew	Leibniz: Determinist, Theist, Idealist - Connections to Spinoza
Alexander Samuel	Space, Time, and Deity - A Study in Panpsychism
Allison Henry E.	Benedict de Spinoza - An Introduction
Ameriks Karl	Kant and the Fate of Spinozism
Andler Charles	Spinoza et ses contemporains
Angelelli Ignacio	Studies on Panpsychism and Consciousness
Arndt Andreas	Spinoza in der deutschen Philosophie
Arnold Matthew	Essays in Criticism - Reflections on Spinoza
Ash Matthew	Panpsychism and the Mind-Body Problem
Aubenque Pierre	Spinoza et le problème du salut
Audi Robert	A Contemporary Perspective on Panpsychism
Avenarius Richard	Kritik der reinen Erfahrung - Towards Panpsychism
Baird Robert M.	Spinoza: Selections and Commentary
Baker Gordon	Wittgenstein and Spinoza - Ethics and Logic
Balibar Étienne	Spinoza et la politique
Barkley Stephen	Panpsychism and Physical Theory
Barrett William	Irrational Man - Encounters with Spinoza and Existentialism
Bayle Pierre	Dictionnaire historique et critique - Spinoza entry
Beauchamp Tom L.	Philosophical Ethics - Including Spinoza's Views
Beiser Frederick C.	German Idealism and Spinoza
Bell Daniel	Spinoza in Twenty-First Century Philosophy
Bennett Jonathan	A Study of Spinoza's Ethics
Bentley Eric	Mind in Nature - A Panpsychist Approach
Berman David	Consciousness and Panpsychism

Bernstein Richard J.	Spinoza, Ethics, and Modernity
Beth Éwald	La philosophie de Spinoza
Blaustein Albert	Spinoza and American Thought
Blumenberg Hans	Die Lesbarkeit der Welt - Spinoza Influences
Böhme Gernot	Natural Consciousness and Panpsychism
Bohr Niels	Physics and Philosophy - Spinoza Reflections
Bolton Martha	Spinoza and Early Modern Rationalism
Bossert Walter	Panpsychism: Mind Everywhere
Boulnois Olivier	Spinoza et la critique des passions
Bourdon Francis	Le panpsychisme dans l'histoire de la philosophie
Bowen Peter	Panpsychism in Modern Science
Brandon Robert	Spinoza's Metaphysics
Brett George Sydney	The Philosophy of Spinoza
Broad C.D.	Mind and its Place in Nature - Panpsychism Theory
Brook Andrew	Spinoza and the Theory of Mind
Broughton Janet	Descartes and Spinoza
Brown Deborah J.	Spinoza on Reason and the Passions
Bruntrup Godehard	Panpsychism - Contemporary Perspectives
Bunge Mario	Matter and Mind - A Philosophical Inquiry
Burghardt Hans	Spinoza und die Aufklärung
Butterfield Herbert	Spinoza and Enlightenment Politics
Cairns Angus	Panpsychism and the Unity of Nature
Campa Amanda	Spinoza and the Religious Mind
Carlisle Clare	Philosopher of the Infinite - Spinoza
Carr Stephen	Mind in Every Stone - Panpsychism Today
Carter William	Reality and Consciousness - Panpsychism Approach
Cassirer Ernst	Die Philosophie der symbolischen Formen - Spinoza Influence
Chalmers David J.	The Conscious Mind - Panpsychism Perspectives
Chaui Marilena	Politica e filosofia in Spinoza
Chevalier Jean	Spinoza et le spinozisme
Clarke Desmond	Spinoza and Philosophy of Religion
Clements Keith	Panpsychism and Theology
Cohen Herman	Spinoza und die deutsche Philosophie

Coleman William	Panpsychism in Evolutionary Theory
Collingwood R.G.	The Idea of Nature - Panpsychism Elements
Condorcet Jean Antoine	Réflexions sur Spinoza
Connell Michael	Panpsychism and Neuroscience
Cook J. Thomas	Spinoza's Metaphysical Vocabulary
Copleston Frederick	History of Philosophy - Spinoza Volume
Cornford Francis	Plato's Cosmology and Panpsychism
Cottingham John	Spinoza and Modern Ethics
Curley Edwin	Behind the Geometrical Method - Spinoza's Ethics
Damiani Giovanni	Spinoza e l'idea di libertà
Darby George	panpsychism and Physics - A Study
Darwin Charles	Correspondences - Panpsychism and Nature
Della Rocca Michael	Spinoza - A Guide for the Perplexed
Deleuze Gilles	Spinoza et le problème de l'expression
Dennett Daniel C.	Consciousness Explained - Panpsychism Commentaries
Descartes René	Spinoza et Descartes - Correspondances
Devaux André	Spinoza et la modernité
Dietrich Eric	Panpsychism and the Philosophy of Mind
Di Giovanni George	Spinoza in Kant's Age
Dooremalen Hans	Panpsychism: Mind Across Matter
Doyle Sean	Spinoza and Religion
Droysen Johann Gustav	Spinoza in der deutschen Geschichtsschreibung
Duncan Graeme	Panpsychism and Social Theory
Dunham Jeremy	The Rationalists - Spinoza Chapter
Durt Christoph	Mind in the World - Panpsychism Today
Eckstein Arthur	Spinoza's Political Treatises
Edelman Gerald	Neural Darwinism and Panpsychism
Edwards Paul	Spinoza - His Life and Philosophy
Efraim Jason	Panpsychism and Cognitive Science

Elwes Thomas	The Works of Spinoza
Emery Thomas	Panpsychism and Realism
Engels Friedrich	Dialectics of Nature - Panpsychist Passages
Etherton Joel	Panpsychism in Artificial Intelligence Research
Evans James	Spinoza - The Universal Philosopher
Falkenberg Friedrich	Leibniz und Spinoza - Ein Vergleich
Feuerbach Ludwig	Thoughts on Spinoza
Fisch Max H.	Spinoza and Pragmatism
Flanagan Owen	Consciousness Reconsidered - Panpsychism Input
Freudenthal Jacob	Spinoza - Sein Leben und seine Lehre
Friedman Michael	Mind in a Monistic Universe - Panpsychism
Friedman Russell	Spinoza and Medieval Philosophy
Frolov Ivan	Panpsychism in Soviet Thought
Fuller Steve	Spinoza and the Sociology of Knowledge
Giberman Daniel	Contemporary Panpsychism - Bridging Mind and Matter
Gilson Etienne	Études sur la pensée philosophique de Spinoza
Girard René	Spinoza et la violence du réel
Gochet Paul	La philosophie de l'esprit et le panpsychisme
Goldstein Rebecca	Betraying Spinoza - The Renegade Jew Who Gave Us Modernity
Gonthier Jean	Le panpsychisme selon Fechner
Goudriaan Aza	Spinoza and Theology in the Dutch Republic
Grene Marjorie	Approaches to a Philosophical Biology - Panpsychist Elements
Griffin David Ray	Unsnarling the World-Knot - Consciousness, Freedom, and the Mind-Body Problem
Gueroult Martial	Spinoza I: Dieu - Spinoza II: L'âme
Gutting Gary	Spinoza: A Very Short Introduction
Haack Susan	Philosophy of Mind - Panpsychism Considered
Hadot Pierre	Exercices spirituels et philosophie antique - Spinoza Connections
Hampshire Stuart	Spinoza and Paradox
Harrison Paul	Panpsychism and the Evolution of Consciousness
Hart W.D.	The Engines of the Soul - Panpsychism in Perspective
Harrison John	Spinoza and the Philosophical Imagination
Hassan Ihab	Spinoza and the Postmodern Condition
Heath William	Mind and Cosmos - Panpsychism and Science

Heinz Marion	Spinoza and German Philosophy
Heller Agnes	A Philosophy of History from Spinoza to Marx
Hershenov David	Panpsychism and Ethics
Heyd David	Supererogation and Spinoza's Moral Philosophy
Hobbes Thomas	Dialogues with Spinoza
Hodgson Shadworth H.	Philosophy of Reflection: Panpsychism Studies
Hoenen Maarten J.F.M.	Spinoza in Medieval Contexts
Hoffman Paul	Essays on Descartes and Spinoza
Holland Eugene	Deleuze and Spinoza: Expressionism in Philosophy
Holt Jim	Why Does the World Exist? - Panpsychism Hypotheses
Honderich Ted	The Oxford Companion to Philosophy - Panpsychism Entry
Hösle Vittorio	Spinoza und sein Jahrhundert
Houlgate Stephen	Spinoza in German Idealism
Huenemann Charles	Spinoza's Ethics: An Introduction
Hunt Harold	Panpsychism in Contemporary Philosophy
Husserl Edmund	Spinoza und die transzendentale Phänomenologie
Israel Jonathan	Radical Enlightenment - Philosophy and the Making of Modernity (Spinoza)
Jacobs Michael	Panpsychism and Ecology
Jacobs Wilhelm	Spinoza und das deutsche Denken
James William	Essays in Radical Empiricism - Panpsychist Chapters
Jarrett Michael	Spinoza and the Modern Mind
Jasper Karl	Spinoza, Nietzsche, Heidegger
Jaspers Karl	Die großen Philosophen - Spinoza Volume
Jean Léo	Le panpsychisme et la métaphysique de la nature
Joachim Harold H.	A Study of the Ethics of Spinoza
Johnson David	Consciousness and Panpsychism in Analytic Philosophy
Jonas Hans	The Concept of God After Auschwitz - Spinoza Section
Jones Bernard	The Universe and Consciousness - Panpsychism Debate
Jonathan Israel	Enlightenment Contested - Spinoza's Influence
Jung Carl Gustav	Spinoza und die archetypische Symbolik

Kantorowicz Ernst H.	The King's Two Bodies - Spinoza References
Kaplan Benjamin	Spinoza and Religion in the Dutch Golden Age
Kastrup Bernardo	The Idea of the World - A Multi-disciplinary Argument for Panpsychism
Kaufmann Felix	Spinoza und die Philosophie der Neuzeit
Kavanaugh William	Panpsychism and Theological Reflections
Keijzer Fred	Spinoza and Cognitive Science
Kelly Michael	Panpsychism in Philosophy of Science
Kemp Peter	Spinoza et la pensée moderne
Keppler Joseph	Panpsychism and Mysticism
Kerr-Lawson Angus	Spinoza's Theory of Religion
Kivy Peter	The Philosophy of Music and Spinoza
Koistinen Olli-Pekka	The Cambridge Companion to Spinoza
Krämer Hans	Spinoza und die Renaissance des Monismus
Kripke Saul	Spinoza and the Philosophy of Language
Kroes Bert	Panpsychism in Quantum Physics
Kropotkin Peter	Mutual Aid - Panpsychist Remarks
Kuhn Thomas S.	The Structure of Scientific Revolutions - Spinoza in Context
Lakoff George	Philosophy in the Flesh - Spinoza References
Lambert Gregg	In Search of a New Image of Thought - Spinoza and Deleuze
Lange Frederick	History of Materialism - Panpsychism Discussions
Lanfredini Tommaso	Spinoza tra metafisica e politica
Laplace Pierre-Simon	Essai philosophique sur les probabilités - Panpsychism Implications
Larkin Basil	Panpsychism: Spirit in Matter
Laruelle François	Le principe de minorité - Spinoza Influence
Latour Bruno	Face à Gaïa - Panpsychism and Actor-Network Theory
Lauriola Angela	L'etica di Spinoza
Law David	Spinoza's Political Philosophy
Leibniz Gottfried Wilhelm	Monadology - Panpsychist Principles
Leiser Madeline	Panpsychism and the Soul of the World
Lellouch Jacques	Spinoza, lecteur d'Euclide

Lemos Ramon	Panpsychism and Religious Consciousness
Lenin Vladimir I.	Materialism and Empirio-Criticism - Panpsychism Remarks
Lenoir Frédéric	Spinoza, le désir de Dieu
Leslie John	The Universes - Panpsychism Conceptualizations
Levi Primo	Conversazioni e riflessioni su Spinoza
Lewis David	On the Plurality of Worlds - Panpsychism References
Linschoten Johannes	Spinoza - Studies in Dutch Philosophy
Locke John	Spinoza and Locke - Comparative Studies
Lodge Rupert	Philosophy of Loyalty - Panpsychism Notes
Lombardo Giovanni	Panpsichismo e ontologia contemporanea
Longuenesse Béatrice	Spinoza et Kant: Dialectique et liberté
Lorenz Konrad	L'anello di Re Salomone - Panpsychism in Biology
Lovejoy Arthur O.	The Great Chain of Being - Panpsychism Roots
Lowe E.J.	The Possibility of Metaphysics - Panpsychism Relevance
Lucretius Titus	Spinoza and the Ancient Materialists
Luhmann Niklas	Die Gesellschaft der Gesellschaft - Spinoza Citations
Lund Humphrey	Panpsychism and the Philosophy of Life
Lurjia Aleksandr	Neuropsychology and Panpsychism
Lyotard Jean-François	La différend - Spinoza References
Machiavelli Niccolò	Spinoza e il pensiero politico moderno
Mackenzie Matthew	Embodied Cognition and Panpsychism
Macpherson C.B.	Spinoza's Theory of Democracy
Madden Edward H.	Concepts of Nature - Panpsychism Context
Magee Bryan	The Story of Philosophy - Spinoza Section
Maimon Salomon	Spinoza und die Kabbala
Malebranche Nicolas	Controverses avec Spinoza
Manzini Giancarlo	Panpsichismo e scienza della mente
Marcucci Stefano	L'Etica di Spinoza commentata

Margenau Henry	The Nature of Physical Reality - Panpsychism Discourses
Maritain Jacques	Trois Réformateurs - Luther, Descartes, Spinoza
Marshall Colin	Panpsychism and the Emergence of Mind
Martinetti Piero	Spinoza e l'eternità
Marx Karl	Economic and Philosophic Manuscripts - Spinoza Links
Maslin Keith	An Introduction to Philosophy - Spinoza Notes
Massumi Brian	Parables for the Virtual - Spinoza Inspiration
Mathews Freya	For Love of Matter - A Contemporary Panpsychism
McCullough Laurence	Spinoza and Moral Philosophy
McGinn Colin	The Problem of Consciousness - Panpsychism View
McKenna Anthony	Spinoza and the Religious Enlightenment
McTaggart J.M.E.	The Nature of Existence - Panpsychism Part
Meijer Rembrandt	Spinoza en Nederland
Melamed Yitzhak Y.	Spinoza's Metaphysics
Mellor D.H.	The Facts of Causation - Panpsychism Commentary
Mengaldo Pier Vincenzo	Spinoza e la lingua del XVII secolo
Merleau-Ponty Maurice	Le visible et l'invisible - Panpsychism Angle
Merrick Thomas	Panpsychism and the Nature of Consciousness
Mill John Stuart	Essays on Spinoza and Utilitarianism
Mitchell Bruce	Panpsychism and Ancient Philosophy
Mondolfo Rodolfo	Spinoza e la filosofia religiosa
Montero Barbara	Panpsychism and Physicalism
Moore Adrian	The Infinite - Spinoza Chapter
Morin Edgar	La Méthode - Panpsychism within Complexity
Morrison James	Spinoza and Philosophy of Mind
Moschini Fabrizio	Spinoza e le radici dell'illuminismo europeo
Moss Donald	Panpsychism and the Dimensions of the Self
Mounce Howard O.	Spinoza and the Philosophy of Language
Muffato Alberto	Panpsichismo e cosmologia contemporanea
Nagel Thomas	Mind and Cosmos - Arguments Relevant to Panpsychism

Nadler Steven	Spinoza: A Life
Nadler Steven	Spinoza's Ethics: An Introduction
Naitza Enrico	Spinoza e il pensiero scientifico
Nancy Jean-Luc	L'Expérience de la liberté - Riflessioni spinoziane
Nehamas Alexander	Spinoza and the Question of Interpretation
Neilson William Allan	Panpsychism in Modern Thought
Newton Isaac	Spinoza and Natural Philosophy
Nicholson Emily	Panpsychism and Environmental Philosophy
Nietzsche Friedrich	Spinoza und die Wissenschaft vom Glück
Noë Alva	Out of Our Heads - Panpsychism and the Limits of Neuroscience
Noreña Carlos	Spinoza and the Rise of Liberalism
Northcote Thomas	Panpsychism in Theology and Metaphysics
Novak David	Spinoza and Jewish Theology
Oakes Robert	Consciousness and the Universe - Panpsychism Essays
Oakes Robert	The Wisdom of Spinoza
Oberhausen Michael	Spinoza und der deutsche Idealismus
Ocampo Sebastian	Panpsychism and Mind-Matter Unity
Olivetti Cesare	Spinoza e l'ontologia monistica
Ollman Bertell	Dialectical Investigations - Spinoza Insights
Olsen Henry	Panpsychism and Moral Philosophy
Orsucci Andrea	Spinoza e i Lumi italiani
Ortega y Gasset José	Ideas y creencias - Spinoza Rileggibile
Otto Rudolf	God and the Moral Order - Spinoza Considerations
Oxford Henry	Panpsychism in Contemporary Metaphysics
Paci Enzo	Spinoza e il problema dell'infinito
Pagano Leonardo	Il panpsichismo e l'etica naturale
Paggini Luca	Spinoza e la geometria della mente
Palmer Anthony	Panpsychism and Cognitive Science Today
Panaccio Claude	Spinoza et la notion de substance
Parfit Derek	Reasons and Persons - Panpsychist Interpretations
Parker Kelly	The Continuum of Mind - Panpsychism
Parlato Michele	Spinoza e il pensiero radicale
Parodi Marco	Panpsichismo e filosofia della mente

Parsons Charles	Spinoza and the Foundations of Logic
Passmore John	The Perfectibility of Man - Spinoza Sections
Pasteur Louis	Spinoza e la filosofia della natura
Paton H.J.	The Modern Predicament - Panpsychism
Pavese Carlo	Spinoza e la libertà razionale
Peirce Charles Sanders	Collected Papers - Panpsychist Arguments
Pekar Adam	Panpsychism and the Ethics of Life
Penrose Roger	The Emperor's New Mind - Panpsychism References
Péra Lucien	Spinoza et l'humanisme radical
Perkins Franklin	Spinoza as Educator
Perry Ralph Barton	Present Philosophical Tendencies - Panpsychism Chapters
Pesce Domenico	Spinoza e la teologia filosofica
Petit Jean-Luc	Panpsychism and Phenomenology
Petrosino Silvano	Spinoza e il linguaggio dell'eternità
Phillips Dewi	Spinoza and the Philosophy of Religion
Pierucci Antonio	Panpsichismo e filosofia biblica
Pigliaru Antonio	Spinoza e la filosofia del diritto
Pinkard Terry	German Philosophy 1760–1860 - Spinoza Context
Plantinga Alvin	Panpsychism and Theistic Metaphysics
Plazzi Gianfranco	Spinoza e la metafisica della sostanza
Poggi Gianfranco	Spinoza e il pensiero politico europeo
Polkinghorne John	The Mind of God - Panpsychism Perspectives
Pollock John	Spinoza: His Life and Philosophy
Popper Karl	Conjectures and Refutations - Panpsychism Passages
Porchat Olivia	Panpsychism and the Nature of Being
Porro Giovanni	Spinoza e la filosofia delle cause
Portmann Adolf	La forma vivente - Panpsychism in Biology
Possenti Vittorio	Spinoza e la filosofia critica
Prajapati Mohan	Panpsychism across Philosophical Traditions
Prigogine Ilya	The End of Certainty - Panpsychism within Science
Putnam Hilary	Spinoza and the Problem of Representation
Quine Willard Van Orman	From a Logical Point of View - Panpsychism References

Quinton Anthony	The Nature of Things - Panpsychism Views
Quispel Gilles	Spinoza et les sources chrétiennes
Rabbi Yehuda	Spinoza and Jewish Mysticism
Radford Colin	Human Nature - Panpsychism in Ethics
Ragghianti Carlo Ludovico	Spinoza e le arti visive
Ramachandran V.S.	The Tell-Tale Brain - Panpsychism and Consciousness
Ranchio Marco	Panpsichismo e scienza cognitiva
Raphael D.D.	Spinoza and Political Theory
Rasch Thomas	Panpsychism and Cosmology
Rashid Salman	Panpsychism and the Unity of Experience
Rasmussen Dennis	The Pragmatic Enlightenment - Spinoza's Footprint
Reid Thomas	Essays on the Intellectual Powers - Panpsychism
Rescher Nicholas	Process Philosophy - Panpsychism Chapter
Ricœur Paul	Le conflit des interprétations - Spinoza Passages
Ridder Hans	Panpsychism and Evolution
Riemann Bernhard	Spinoza and Mathematical Philosophy
Riva Franco	Spinoza e il pensiero dell'unità
Robinson Daniel	Consciousness and Its Implications - Panpsychism
Rosenstock Bruce	Spinoza, Marrano of Reason
Rossi Paolo	Spinoza e la nascita della scienza moderna
Rovelli Carlo	La realtà non è come appare - Panpsychism in Physics
Russell Bertrand	The Analysis of Mind - Panpsychism Essentials
Ryle Gilbert	Collected Papers - Comments on Spinoza
Sacks Oliver	The River of Consciousness - Panpsychism Perspective
Schaeffer Jean-Marie	Spinoza et les arts
Schaffer Jonathan	Spinoza and Priority Monism
Scheler Max	The Nature of Sympathy - Panpsychism Thought
Scholem Gershom	Major Trends in Jewish Mysticism - Spinoza Section
Searle John	Mind - Panpsychism Critique
Segré Paolo	Spinoza e la filosofia della scienza
Sellars Wilfrid	Science and Metaphysics - Panpsychism
Shapiro Michael	Spinoza and Modern Political Philosophy

Sheldrake Rupert	The Science Delusion - Panpsychism in Biology
Shook John R.	Panpsychism in the History of Philosophy
Simondon Gilbert	Individuation - Spinozist Influence
Singer Peter	Practical Ethics - Spinoza References
Siorvanes Lucas	Panpsychism and Neoplatonism
Skorupski John	Ethics and the Limits of Philosophy - Spinoza Remarks
Smart J.J.C.	Philosophy and Scientific Realism - Panpsychism Input
Smith Steven	Spinoza and the Political
Snow Cecil	Panpsychism and Education
Sosa Ernest	Knowledge in Perspective - Spinoza Discussions
Sovati David	Panpsychism in Contemporary Ontology
Speranza Gabriele	Spinoza e il concetto di verità
Sprigge Timothy L.S.	The God of Metaphysics - Panpsychism Argument
Staten Henry	Spinoza: the Enthralling Philosophy
Steiner George	Spinoza in Our Time
Sterckx Frank	Panpsychism and Consciousness Studies
Stock Mauro	Spinoza e la modernità radicale
Stone T.D.	Panpsychism and Metaphysical Realism
Stout Rowland	Philosophy of Mind - Panpsychism Notes
Strawson Galen	Realistic Monism - Panpsychism
Struthers Wendy	Spinoza and Feminist Philosophy
Suffritti Lorenzo	Panpsichismo e psicologia filosofica
Sullivan Robert	Spinoza's Political Theory
Swedenborg Emanuel	Spinoza and Mystical Theology
Tavani Giorgio	Panpsichismo e filosofia naturale
Taylor Charles	Sources of the Self - Spinoza Context
Teilhard de Chardin Pierre	The Phenomenon of Man - Panpsychism Thought
Thiel Udo	Spinoza and Early Modern Philosophy of Mind
Thomas William	Panpsychism and Artificial Intelligence
Thomson Iain	Heidegger and Spinoza
Timpanaro Sebastiano	Spinoza e la filosofia materialista
Tolstoj Lev	Spinoza and the Spirit of Freedom
Tonelli Giorgio	Studi su Spinoza
Torres Juan	Panpsychism in Philosophy of Religion

Toto Claudio	Spinoza e filosofia della percezione
Toulmin Stephen	Cosmopolis - Spinoza in Modernity
Traldi Marco	Panpsichismo ed epistemologia
Tricaud Franck	Spinoza et la raison pratique
Turro Sebastian	Panpsychism and Cognitive Theories
Underhill Evelyn	Spinoza and Mystical Philosophy
Urbani Giovanni	Panpsichismo e scienze umane
Valladar Marcos	Spinoza e il pensiero critico
Vallee Jacques	Dimensions - Panpsychism in Science
Van Cleve James	Philosophy of Mind and Matter - Panpsychism
Vanini Giulio Cesare	Spinoza e il libero pensiero
Vazquez Daniel	Panpsychism and the Structure of Reality
Verdiglione Armando	Spinoza e la parola
Verhaeghe Paul	Panpsychism and Psychoanalysis
Verhoeven Cornelius	Spinoza - Filosofia della speranza
Vianna Ricardo	Panpsychism and Physical Reality
Villanueva Enrique	Spinoza et le monde contemporain
Vinci Joseph	Panpsychism and Rationality
Vinti Antonio	Spinoza e la filosofia italiana
Vitalini Mauro	Panpsi-chismo e scienza della coscienza
Voegelin Eric	The New Science of Politics - Spinoza Context
Voltaire François-Marie	Lettres philosophiques - Spinoza Citations
Von Wright Georg Henrik	Philosophical Logic - Panpsychism Notes
Waldron Jeremy	Spinoza and Modern Constitutionalism
Walker Ralph	Panpsychism and the Concept of Mind
Walther Manfred	Spinoza und die deutsche Klassik
Ward Andrew	Rational Theology and Panpsychism
Watson Richard	The Downfall of Cartesianism - Spinoza as Alternative
Weber Max	Spinoza and Sociology
Weiss Paul	Modes of Being - Panpsychism
White Alan	Spinoza and the Philosophy of Freedom

Whitehead Alfred North	Process and Reality - Panpsychism Framework
Whitney Charles	Panpsychism and Literary Philosophy
Wiggins David	Needs, Values, Truth - Spinoza Comments
Williams Bernard	Ethics and the Limits of Philosophy - Spinoza Insights
Wilson Margaret	Ideas and the Mind - Panpsychism Link
Wittgenstein Ludwig	Remarks on Spinoza
Wolff Christian	Spinoza und die deutsche Metaphysik
Woolhouse Roger	Spinoza's Theory of Knowledge
Worrall John	Panpsychism and the Philosophy of Science
Yaffe Gideon	Liberty Worth the Name - Spinoza's Theory
Young Robert	Panpsychism - Mind Throughout Nature
Zaehner R.C.	Mysticism, Sacred and Profane - Spinoza Context
Zaccaria Giuseppe	Spinoza e la filosofia del diritto contemporaneo
Zagzebski Linda	Panpsychism and Virtue Ethics
Zammito John	A Nice Derangement of Epistemes - Spinoza Footnotes
Zanetti Giovanni	Panpsichismo e ontologia
Zarader Jean-Pierre	Spinoza et la philosophie du XVIIe siècle
Zeller Eduard	Geschichte der Philosophie - Spinoza Section
Zilsel Edgar	Spinoza and Science
Zimmerman Dean W.	Metaphysics and Panpsychism
Zini Sergio	Spinoza e l'educazione etica
Zippel Marco	Panpsichismo e neuroscienze
Zubiri Xavier	Nature, History, God - Panpsychism in Metaphysics
Zuccarini Paolo	Spinoza e l'arte della ragione
Zunini Luigi	Panpsichismo e logica filosofica

What if the universe were alive? In stone, in the wind, in distant stars: everywhere, a silent echo of consciousness.

Panpsychism—the idea that the mind is not a privilege of humans but a fundamental property of reality—spans millennia of thought, from archaic myths to the most advanced scientific laboratories.

In these pages, the journey begins with the wisdom of the origins, passes through ancient and medieval philosophers, delves into the work of Spinoza and his radical Deus sive Natura, and then continues through the Enlightenment, modern philosophy, and the most recent hypotheses on consciousness, including bold quantum theories.

With a perspective that blends rigor and imagination, this book explores the ethical and metaphysical implications of a worldview in which everything is interconnected: where understanding Nature means rediscovering the divine in the very breath of reality.

An invitation to rethink what we call 'life', 'mind' and 'matter', to open the door to a new perception of reality.

"A cultured and compelling journey into the heart of universal thought: this book has opened up perspectives I was unaware of, making panpsychism a living key to understanding reality."

"A rare balance between academic rigor and narrative skill. I felt like I was walking between ancient myths and modern scientific speculations, without ever losing the thread."

"Finally, a work that addresses the theme of consciousness not as an unsolvable mystery, but as an integral part of the universe. An enlightening and stimulating read."

"I have read many complex essays, but here I found clarity and poetry. Each chapter is a window onto the millennial dialogue between man and nature."

"The concept of Deus sive Natura takes on new life through these pages. A book that makes you think, and perhaps even transforms the way you look at the

The author.

Bruno Del Medico, born in 1946 in Sabaudia (Italy), worked in computer science and software development during his professional life. He later devoted himself to the study and dissemination of the most current scientific theories, such as quantum theory in its metaphysical sense. Bruno Del Medico is currently the author of numerous works published by Edizioni PensareDiverso. Bruno Del Medico's work stands out from traditional narratives in the field of technical sciences for its critical, independent, and unconventional approach. Instead of limiting himself to standard interpretations or general consensus on purely scientific issues, Bruno Del Medico explores alternative perspectives that often go against the grain, questioning commonly accepted assumptions. A distinctive feature of his works is his ability to weave science, ethics, and philosophy into an open dialogue that encourages readers to think for themselves and consider the deeper implications of technological and scientific innovations.

entanglement quantum mechanical physics,
Panpsychism – pantheism – Baruch Spinoza,
mind brain consciousness - quantum consciousness,
philosophy of science - metaphysics,
Penrose - Hameroff,
David Chalmers – hard problem ,
Thomas Nagel - Philip Goff,

www.ingramcontent.com/pod-product-compliance
Lightning Source LLC
Chambersburg PA
CBHW071414230926
51846CB00001B/85
* 9 7 8 8 8 3 2 2 7 3 9 1 5 *